Numerical Methods for Stiff Equations

and Singular Perturbation Problems

Mathematics and Its Applications

Managing Editor:

M. HAZEWINKEL
Department of Mathematics, Erasmus University, Rotterdam, The Netherlands

Volume 5

Willard L. Miranker

*Mathematical Sciences Dept., IBM, Thomas J. Watson Research Center,
Yorktown Heights, N.Y., U.S.A.*

Numerical Methods for Stiff Equations

and Singular Perturbation Problems

D. REIDEL PUBLISHING COMPANY

Dordrecht : Holland / Boston : U.S.A. / London : England

Library of Congress Cataloging in Publication Data

Miranker, Willard L.
    Numerical methods for stiff equations and singular perturbation problems.

    (Mathematics and its applications; V. 5)
    Bibliography: P.
    1. Differential equations—Numerical solutions. 2. Perturbation (Mathematics).
I. Title. II. Series: Mathematics and its applications (Dordrecht); V.5.
QA371.M58   515.3'5   80-13085

ISBN 1-4020-0298-X
Transferred to Digital Print 2001

Published by D. Reidel Publishing Company
P.O. Box 17,3300 AA Dordrecht, Holland

Sold and distributed in the U.S.A. and Canada
by Kluwer Boston Inc.,
190 Old Derby Street, Hingham, MA 02043, U.S.A.

In all other countries, sold and distributed
by Kluwer Academic Publishers Group,
P.O. Box 322, 3300 AH Dordrecht, Holland

D. Reidel Publishing Company is a member of the Kluwer Group

Editor's Preface

Approach your problems from
the right end and begin with the
answers. Then, one day, perhaps
you will find the final question.
'The Hermit Clad in Crane Feathers'
in R. Van Gulik's *The Chinese Maze
Murders.*

It isn't that they can't see the
solution. It is that they can't see
the problem.

G. K. Chesterton, The scandal of
Father Brown "The point of a pin"

Growing specialization and diversification have brought a host of
monographs and textbooks on increasingly specialized topics. However,
the 'tree' of knowledge of mathematics and related fields does not grow
only by putting forth new branches. It also happens, quite often in fact,
that branches which were thought to be completely disparate are suddenly
seen to be related.

Further, the kind and level of sophistication of mathematics applied in
various sciences has changed drastically in recent years: measure theory
is used (non-trivially) in regional and theoretical economics; algebraic
geometry interacts with physics; the Minkowsky lemma, coding theory
and the structure of water meet one another in packing and covering
theory; quantum fields, crystal defects and mathematical programming
profit from homotopy theory; Lie algebras are relevant to filtering; and
prediction and electrical engineering can use Stein spaces.

This series of books. *Mathematics and Its Applications*, is devoted to such
(new) interrelations as *exempla gratia*:

- -a central concept which plays an important role in several different
  mathematical and/or scientific specialized areas;
- —new applications of the results and ideas from one area of scientific
  endeavor into another;
- —influences which the results, problems and concepts of one field of
  enquiry have and have had on the development of another.

With books on topics such as these, of moderate length and price, which
are stimulating rather than definitive, intriguing rather than encyclo-
paedic, we hope to contribute something towards better communication
among the practitioners in diversified fields.

The unreasonable effectiveness of
mathematics in science . . .

　　　　　　Eugene Wigner

Well, if you knows of a better 'ole,
go to it.

　　　　Bruce Bairnsfather

What is now proved was once only
imagined.

　　　　　　William Blake

As long as algebra and geometry
proceeded a long separate paths,
their advance was slow and their
applications limited.

　　But when these sciences joined
company, they drew from each
other fresh vitality and thence-
forward marched on at a rapid pace
towards perfection.

　　　　Joseph Louis Lagrange

Krimpen a/d IJssel
March, 1979.

MICHIEL HAZEWINKEL

Table of Contents

Preface

Two principal approaches to the problems of applied mathematics are through numerical analysis and perturbation theory. In this monograph, we discuss and bring together a special body of techniques from each of these: (i) from numerical analysis, methods for stiff systems of differential equations, (ii) from perturbation theory, singular perturbation methods. Both of these areas are grounded in problems arising in applications from outside of mathematics for the most part. We cite and discuss many of them.

The mathematical problem treated is the initial value problem for a system of ordinary differential equations. However, results for other problems such as recurrences, boundary value problems and the initial value problem for partial differential equations are also included.

Although great advances have by now been made in numerical methods, there are many problems which seriously tax or defy them. Such problems need not be massive or ramified. Some are the simplest problems to state. They are those problems which possess solutions which are particularly sensitive to data changes or correspondingly problems for which small changes in the independent variable lead to large changes in the solution. These problems are variously called ill conditioned, unstable, nearly singular, etc. Stiff differential equations is a term given to describe such behavior for initial value problems.

Problems of this type have always attracted attention among mathematicians. The stiff differential equation is a relative late comer, its tardiness correlated perhaps to the development of powerful computers. However, in recent years a sizeable collection of results has emerged for this problem, although of course very much remains to be done. For example, the connection between stiff problems and other types of ill conditioned problems is easy to draw. However, there is a conspicious paucity of methods of regularization, so commonly used for ill conditioned problems in the treatment of stiff equations.

Problems of singular perturbation type are also ill conditioned in the sense described here. These problems have been extensively and continuously studied for some time. Only relatively recently and also with the development of powerful computers has the numerical analysis of such problems begun in a significant way.

Of course the two problem classes overlap as do the sets of numerical methods for each. We include examples and applications as well as the results of illustrative computational experiments performed with the methods discussed here. We also see that these methods form the starting point for additional numerical study of other kinds of stiff and/or singularly perturbed problems. For this reason, numerical analyses of recurrences, of boundary value problems and of partial differential equations are also included.

Most of the material presented here is drawn from the recent literature. We refer to the survey of Bjurel, Dahlquist, Lindberg and Linde, 1972, to lecture notes of Liniger, 1974 and of Miranker, 1975, to three symposia proceedings, one edited by Willoughby, 1974, one by Hemker and Miller, 1979, and one by Axelsson, Frank and vander Sluis, 1980. These and citations made in the text itself to original sources are collected in the list of references. I cite particularly the work of F. C. Hoppensteadt to whom is due (jointly with myself) all of the multitime methodology which is presented here.

This monograph is an outgrowth of an earlier one which contained my lecture notes for courses given at the Université de Paris-Sud, (Orsay) and at the Instituto per le Applicazioni del Calcolo 'Mauro Picone', Rome during 1974–1975.

The presentation in this monograph reflects the current active state of the subject matter. It varies from formal to informal with many states in between. I believe that this shifting of form is not distracting, but on the contrary, it serves to stimulate understanding by exposing the applied nature of the subject on the one hand and the interesting mathematics on the other. It certainly shows the development of mathematics as a subject drawing on real problems and supplying them in turn with structure, a process of mutual enrichment. This so-called process of applied mathematics is one which I learned so many years ago as a student, first of E. Isaacson and then of J. B. Keller, and I do, with gratitude, dedicate this modest text to them.

I am grateful to R. A. Toupin, the Director of the Mathematical Sciences

Department of the IBM Research Center for his interest in and for his support of this work. For the physical preparation of the text, I must thank Jo Genzano, without whose help I would not have dared to attempt it.

Yorktown Heights, 1979   W. L. M.

Chapter 1

Introduction

Summary

In the first section of this chapter, we introduce the problem classes to be studied in this monograph. In the second section, we review the classical linear multistep theory for the numerical approach to ordinary differential equations.

The problem classes, which as we will see are rather closely related to each other, are stiff differential equations and differential equations of singular perturbation type. Our introduction to them is complemented by the presentation examples both of model problems and of actual applications.

These two problem classes seriously defy traditional numerical methods. The numerical approach to these problems consists of exposing the limitations of the traditional methods and the development of remedies. Thus, we include the review of the linear multistep theory here since it is the traditional numerical theory for differential equations and as such it supplies the point of departure of our subject.

1.1. STIFFNESS AND SINGULAR PERTURBATIONS

1.1.1. *Motivation*

Stiff differential equations are equations which are *ill-conditioned* in a computational sense. To reveal the nature of the ill-conditioning and to motivate the need to study numerical methods for stiff differential equations, let us consider an elementary error analysis for the *initial value problem*

$$\dot{y} = -Ay, \qquad 0 < t \leq \bar{t},$$
$$y(0) = y_0.$$

(1.1.1)

Here y is an m-vector and A is a constant $m \times m$ matrix. The dot denotes time differentiation. Corresponding to the increment $h > 0$, we introduce the *mesh points* $t_n = nh$, $n = 0, 1, \ldots$. The solution

$$y_n = y(t_n),$$

of (1.1.1) obeys the recurrence relation,

$$y_{n+1} = e^{-Ah} y_n. \tag{1.1.2}$$

For convenience we introduce the function $S(z) = e^{-z}$, and we rewrite (1.1.2) as

$$y_{n+1} = S(Ah) y_n. \tag{1.1.3}$$

The simplest numerical procedure for determining an approximation u_n to y_n, $n = 1, 2, \ldots$, is furnished by *Euler's method*,

$$u_{n+1} - u_n = -hA u_n, \qquad n = 1, 2, \ldots, \tag{1.1.4}$$
$$u_0 = y_0.$$

Using the function $K(z) = 1 - z$, we may rewrite (1.1.4) as

$$u_{n+1} = K(Ah) u_n. \tag{1.1.5}$$

$K(z)$ is called the *amplification factor* and $K(Ah)$ the *amplification operator* corresponding to the difference equation (1.1.4).

By subtracting (1.1.5) from (1.1.3), we find that the *global error*,

$$e_n = u_n - y_n,$$

obeys the recurrence relation

$$e_{n+1} = K e_n + T y_n. \tag{1.1.6}$$

Here $T = K - S$ is the *truncation operator*. (1.1.6) may be solved to yield

$$e_{n+1} = \sum_{j=0}^{n} K^j T y_{n-j},$$

from which we obtain the bound

$$\| e_n \| \leq n \max_{0 \leq j \leq n-1} \| K \|^j \max_{0 \leq j \leq n-1} \| T y_j \|. \tag{1.1.7}$$

Note that $nh \leq \bar{t}$. Here and throughout this text (and unless otherwise specified) the double bars, $\| \cdot \|$, denote some vector norm or the associated matrix norm, as the case may be.

If the numerical method is *stable*, i.e.,

$$\| K \| \leq 1 \tag{1.1.8}$$

and *accurate* of *order p*, i.e.,

$$\| Ty \| = O(h^{p+1}) \| y \|, \tag{1.1.9}$$

then the bound (1.1.7) shows that $\| e_n \| = O(h^p)$. (Of course for Euler's method, to which case we restrict ourselves, $p = 1$.)

To demonstrate (1.1.9), we note that $\| y \|$ is bounded as a function of t for $0 \leq t \leq \bar{t}$, and we show that $\| T \| = O(h^2)$. For the latter we use the *spectral representation theorem* which contains the assertion

$$T(hA) = \sum_{j=1}^{m} T(h\lambda_j) P_j(A). \tag{1.1.10}$$

Here we assume that the eigenvalues $\lambda_j, j = 1, \ldots, m$ of A are distinct. The $P_j(z)$, $j = 1, \ldots, m$ are the *fundamental polynomials* on the spectrum of A. (i.e., $P_j(z)$ is the polynomial of minimal degree such that $P_j(\lambda_i) = \delta_{ij}$, $i, j = 1, \ldots, m$. Here δ_{ij} is the Kronecker delta.)

We have chosen $T(z) = K(z) - S(z)$ to be small at a single point, $z = 0$. Indeed

$$T(z) = O(z^2).$$

This and (1.1.10) assures that $\| T \| = O(h^2)$. More precisely we have that

$$\| T \| = O(|\lambda_{max}|^2 h^2), \tag{1.1.11}$$

where

$$|\lambda_{max}| = \max_{1 \leq j \leq m} |\lambda_j|.$$

One proceeds similarly, using the spectral representation theorem to deal with the requirement of stability. For Euler's method we obtain stability if

$$|1 - h\lambda_j| \leq 1, \qquad j = 1, \ldots, m. \tag{1.1.12}$$

(See Definition 1.2.11 and Theorem 1.2.12 below.)

For the usual equations encountered in numerical analysis, $|\lambda_{max}|$ is not too large, and (1.1.12) is achieved with a reasonable restriction on the size of h. In turn (1.1.11) combined with the bound (1.1.7) for $\| e_n \|$ yields an acceptable error size for a reasonable restriction on the size of h.

1.1.2. *Stiffness*

For the time being at least, *stiffness* will be an informal idea.

A *stiff system* of equations is one for which $|\lambda_{max}|$ is enormous, so that either the stability or the error bound or both can only be assured by unreasonable restrictions on h (i.e., an excessively small h requiring too many steps to solve the initial value problem). Enormous means, enormous relative to a scale which here is $1/\bar{t}$. Thus, an equation with $|\lambda_{max}|$ small may also be viewed as stiff if we must solve it for great values of time.

In the literature, stiffness for the system (1.1.1) of differential equations is frequently found to be defined as the case where the ratio of the eigenvalues of A of largest and smallest magnitude, respectively, is large. This definition is unduly restrictive. Indeed as we may see, a single equation can be stiff. Moreover, this usual definition excludes the obviously stiff system corresponding to a high frequency harmonic oscillator, viz.

$$\ddot{y} + \omega^2 y = 0, \qquad \omega^2 \text{ large}. \tag{1.1.13}$$

Indeed neither definition is entirely useful in the nonautonomous or nonlinear case. While stiffness is an informal notion, we can include most of the problems which are of interest by using the idea of *ill-conditioning* (i.e., *instability*). Suppose we develop the numerical approximation to the solution of a differential equation along the points of a mesh, for example, by means of a relation of the type (1.1.5). If small changes in u_n in (1.1.5) result in large changes in u_{n+1}, then the numerical method represented by (1.1.5), when applied to the problem in question, is ill-conditioned. To exclude the case wherein this unstable behavior is caused by the numerical method and is not a difficulty intrinsic to the differential equations, we will say that a system of differential equations is stiff if this unstable behavior occurs in the solutions of the differential equations. More formally we have the following definition.

DEFINITION 1.1.1. A system of differential equations is said to be stiff on the interval $[0, \bar{t}]$, if there exists a solution of that system a component of which has a variation on that interval which is large compared to $1/\bar{t}$.

We make the following observation about the informal nature of our discussion.

REMARK 1.1.2. We may ask what the term 'large compared to' signifies in a formal definition. In fact it has no precise meaning, and we are allowing informal notions (like: reasonable restriction, enormous, acceptable, too many, etc.) with which some numerical analysts feel comfortable to find their way into a formal mathematical statement. While allowing this risks some confusion, we will for reasons of convenience continue to do so. In order to minimize this risk and as a model for similar questions, we now explain how this informality could be repaired in the context of Definition 1.1.1. Hereafter we will not return to this point for other similar problems. The repair is made by replacing single objects by a class of objects out of which the single object is drawn.

For example, a proper alternate to Definition 1.1.1 could be the following.

DEFINITION 1.1.3. A collection of systems of differential equations is said to be stiff on an interval $[0,\bar{t}]$, if there exists no positive constant M such that the variation of every component of every solution of every member of the collection is bounded by M.

The following example shows how treacherous the reliance on eigenvalues to characterize stiffness can be; even in the linear case.

Example

$$\dot{y} = A(t)y, \qquad (1.1.14)$$

where

$$A(t) = \begin{bmatrix} \sin \omega t & \cos \omega t \\ \cos \omega t & -\sin \omega t \end{bmatrix}.$$

The eigenvalues of $A(t)$ are ± 1. The matrizant of (1.1.14) is

$$\Phi(t) = B(t)\frac{\sinh \sigma}{\sigma} + I \cosh \sigma.$$

Here I is the 2×2 identity matrix,

$$\sigma = \sqrt{2}(1 - \cos \omega t)^{1/2}$$

and

$$B(t) = \frac{1}{\omega}\begin{bmatrix} 1 - \cos \omega t & \sin \omega t \\ \sin \omega t & \cos \omega t - 1 \end{bmatrix}.$$

Thus for $\omega \to \infty$,

$$\Phi(t) = (\cosh\sqrt{2 - 2\cos \omega t})(1 + O(\omega^{-1}))I$$

uniformly for $t \in [0,\bar{t}]$.

Thus, in spite of the eigenvalues of $A(t)$, the solution of (1.1.14) varies with frequency ω, a quantity at our disposal.

1.1.3. *Singular Perturbations*

The study of problems of *singular perturbation type* predates the study of numerical methods for stiff differential equations. (Perhaps the first appearance of the latter term is in 1952. See Curtiss and Hirschfelder, 1952.) In any case, it is easy to see that problems of singular perturbation type are a subclass of stiff equations.

Indeed the common picture of *boundary layers* and *outer solutions* characterizing solutions of problems of singular perturbation type is analogous to the *rapidly* and *slowly varying modes* of stiff problems. Observing the connection between these two subjects as well as exploiting the large body of techniques from the older theory seems to have begun only as late as the early 70's.

We note this connection here by pointing out that the generic initial value problem for a singularly perturbed system of differential equations may be written in the following form.

$$dx/dt = f(t, x, y, \varepsilon), \qquad x(0) = \xi,$$
$$\varepsilon \, dy/dt = g(t, x, y, \varepsilon), \qquad y(0) = \eta.$$

f and g depend regularly on ε and $g(t, x, y, 0) = 0$. We may observe that this class of systems is stiff by means of the following scalar example. Take

$$f = y, \qquad g = x + y.$$

The eigenvalues of the corresponding system are

$$\varepsilon^{-1} + O(1) \quad \text{and} \quad -1 + O(\varepsilon).$$

In a sense *the smaller ε, the stiffer the system*. Additional examples will be given in the next section.

Of course, there is considerable interest in computational questions for problems of singular perturbation type independent of their role as a subset of the class of stiff problems. Thus, numerical developments for such problems are of interest in their own right as well as for those cases in which these developments are exploitable for numerical purposes for stiff but not necessarily singularly perturbed problems.

This monograph is organized so that the methods for stiff differential equations are presented first, to be followed by methods for the singular perturbation problems. In actual fact we prefer to deemphasize this division.

1.1.4. *Applications*

It is easy to generate model problems which are stiff or singularly perturbed, since it suffices to take a linear system of differential equations whose coefficient matrix has widely separated eigenvalues. One can find model problems of this type, which are used for illustrative computational purposes, scattered throughout this text.

In this section, we give a sampling of examples which arise in applications. We have chosen these examples from circuit analysis, chemical reactions, engineering applications and even mathematical areas among others.

Tunnel Diodes

Tunnel diodes are employed in many high speed circuits. These circuits are modeled by stiff equations with both rapidly equilibrating solutions on the one hand and highly oscillatory solutions on the other.

A simple circuit representing a tunnel diode is given in Figure 1.1-1a. The current through the nonlinear element in the circuit is given by $I = f(v)$, where the tunnel diode characteristic, $f(v)$, is the *S*-shaped graph as indicated in Figure 1.1-1b. The differential equations describing this circuit are

$$C\frac{dv}{dt} = i - f(v),$$

$$L\frac{di}{dt} = E - Ri - v.$$

(1.1.15)

For certain ranges of values of the parameters, (1.1.15) is a stiff system whose solutions exhibit a variety of extreme behavior. We now give an indication of this behavior.

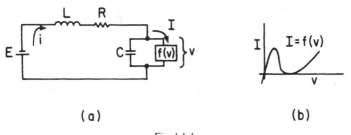

(a)   (b)

Fig. 1.1-1.

Introduce the new variables

$$x = \frac{R}{L}t, \quad \varepsilon = \frac{CR^2}{L}, \quad I = Ri, \quad F(v) = Rf(v)$$

in (1.1.15). We get

$$\varepsilon \frac{dv}{dx} = I - F(v),$$

$$\frac{dI}{dx} = E - v - I.$$

When ε is small this system is stiff, and solutions move alternately through regions of slow change and rapid change. A typical family of solutions in this case is shown in Figure 1.1-2a. The limiting trajectory for ε tending to zero in Figure 1.1-2a, is a curve which has a horizontal portion as well as two parts which coincide with segments of the curve $I = F(v)$ (the tunnel diode characteristic). The changes in I and v per unit x (the dimensionless time) along the latter segments are slow. Correspondingly with respect to the horizontal portion, there is no change in I but a change in V which occurs in zero time. This is typical boundary layer behavior for a singular perturbation problem, i.e., (time) regions of rapid transition. The example fits our notion of a stiff equation as well, since there are large variations of solution in fixed time intervals (the horizontal portion).

Alternatively we may introduce the variables

$$z = t/RC, \qquad \delta = 1/\varepsilon,$$

in terms of which (1.1.15) becomes

$$\frac{dv}{dz} = I - F(v),$$

$$\delta \frac{dI}{dz} = E - I - v.$$

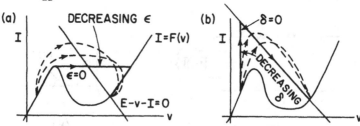

Fig. 1.1-2.

When δ is small, the solutions behave in an extreme manner as shown in Figure 1.1-2b. The behavior for this example with respect to boundary layer or stiff behavior is analogous to the behavior in Figure 1.1-2a. The horizontal portion in the former case has as its counterpart a vertical portion for the latter case. (See Figure 1.1-2.)

A different form of extreme behavior of solutions of (1.1.15) for certain other ranges of parameter values may be observed by introducing the variable

$$y = t/\sqrt{LC}$$

and writing the system (1.1.15) as a single equation

$$\frac{d^2v}{dy^2} + \omega^2 v = K\left(v, \frac{dv}{dy}\right), \tag{1.1.16}$$

where

$$K = (\omega^2 - 1)v - Rf(v) - \left[R\sqrt{\frac{C}{L}} + \sqrt{\frac{L}{C}}f'(v)\right]\frac{dv}{dy} + E.$$

In this case, there are one and sometimes two periodic solutions of (1.1.16) as illustrated in Figure 1.1-3.

These periodic solutions are limit cycles with the approximate frequency

$$\omega = \sqrt{C(L - R^2)}$$

in the t-time scale. Thus for certain values of C, L and R, there are oscillatory solutions of arbitrarily high frequency.

For details concerning asymptotic analyses of the types of solution classes referred to here, see Miranker, 1962 and 1962b.

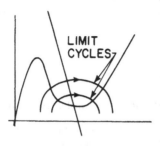

LIMIT
CYCLES

Fig. 1.1-3.

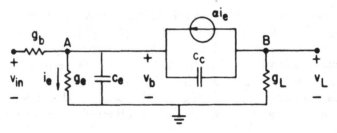

Fig. 1.1-4.

Transistors

A typical equivalent circuit for a transistor is given in Figure 1.1-4.

In terms of the input voltage v_{in}, the voltages $v_b(t)$ and $v_L(t)$ in this circuit are specified by the following differential equations which are statements of Kirchoff's current law at the nodes (A) and (B), respectively.

(A) $g_b v_{in} = (g_b + (1 - \alpha)g_e)v_b + (c_e + c_c)\dfrac{dv_b}{dt} - c_c\dfrac{dv_L}{dt}.$

(B) $0 = \alpha g_e v_b - c_c\dfrac{dv_b}{dt} + g_L v_L + c_C\dfrac{dv_L}{dt}.$

In contemporary transistor technology typical values for the constants appearing here are: $g_e = 0.4$ ohms^{-1}, $g_L = g_b = 0.02$ ohms^{-1}, $\alpha = 0.99$, $c_e = 10^{-9}$ farads and $c_c = 10^{-12}$ farads. In this case the eigenvalues of these differential equations are approximately -2.4×10^7 and -2×10^{10}, respectively. The first of these is the time constant relevant to the response of the transistor which is of physical interest; the second corresponds to an ever-present parasitic effect which makes for increased stiffness in the system.

Thermal Decomposition of Ozone

The kinetic steps involved for a dilute ozone–oxygen mixture are

$$O_3 + O_2 \underset{k_2}{\overset{k_1}{\rightleftharpoons}} O + 2O_2,$$

$$O_3 + O \overset{k_3}{\rightarrow} 2O_2.$$

If the following dimensionless variables are defined:

$$x = [O_3]/[O_3]_0, \quad y = [O]/\varepsilon[O_3]_0,$$
$$\kappa = 2k_2[O_2]_0/k_1, \quad \varepsilon = k_1[O_2]_0/2k_3[O_3]_0,$$

and the time scale divided by $2/k_1[O_2]_0$, then the transient behavior is described by

$$\frac{dx}{dt} = -x - xy + \varepsilon\kappa y, \quad x(0) = 1,$$

$$\varepsilon\frac{dy}{dt} = x - xy - \varepsilon\kappa y, \qquad y(0) = 0.$$

It is interesting to note that this stiff system with $\kappa = O(\varepsilon^{-1})$ corresponds to the simple enzyme reaction treated in Section 5.3 below.

Behavior of a Catalytic Fluidized Bed

Mass and heat balance equations corresponding to one irreversible gas phase reaction within a uniform porous catalyst generate a stiff system of differential equations. For certain numerically specified values of the parameters, such a system is

$$\frac{dx}{dt} = 1.30(y_2 - x) + 1.04 \times 10^4 ky_1, \qquad x(0) = 759,$$

$$\frac{dy_1}{dt} = 1.88 \times 10^3[y_3 - y_1(1 + k)], \qquad y_1(0) = 0.0,$$

$$\frac{dy_2}{dt} = 1752 - 269y_2 + 267x, \qquad y_2(0) = 600,$$

$$\frac{dy_3}{dt} = 0.1 + 320y_1 - 321y_3, \qquad y_3(0) = 0.1,$$

where $k = 0.006 \exp(20.7 - 15000/x)$ and x, y_1, y_2, and y_3 represent the temperature and partial pressure of the catalyst and interstitial fluid, respectively.

For further details concerning these two applications including illustrative computations, we refer to Aiken and Lapidus, 1974.

Recurrence Relations

In Section 7.1, we consider stiff and singularly perturbed recurrence relations. Such relations arise in several areas of application which are discussed in Section 7.1. These are a training algorithm arising in pattern recognition, a population genetics model and regression analysis:

Training Algorithm
Referring to Section 7.1.1, for fixed $\theta > 0$ and $w_0 \in R^{p+1}$, we define $w_1, w_2 \ldots$, as follows.

$$w_{n+1} = w_n + x_n S\left(\frac{w_n \cdot x_n}{\theta}; x_n\right),$$

where

$$S\left(\frac{w \cdot x}{\theta}; x\right) = \begin{cases} \left.\begin{array}{l} 1, w \cdot x \leq \theta \\ 0, w \cdot x > \theta \end{array}\right\} \text{ and } x \in A^*, \\[20pt] \left.\begin{array}{l} -1, w \cdot x \geq -\theta \\ 0, w \cdot x > -\theta \end{array}\right\} \text{ and } x \in B^*. \end{cases}$$

Here A^* and B^* are specified finite point sets in R^{p+1}. Setting $w_n = \theta z(n)$ and $\varepsilon = \theta^{-1}$, the recurrence relation for w_n becomes

$$z(n + 1) = z(n) + \varepsilon x_n S(z(n) \cdot x_n; x_n),$$

while in the definition of S, w is changed to z and θ is replaced by unity.

A Population Genetics Model
In a large population of diploid organisms having discrete generations, the genotypes determined by one locus having two alleles, A and a, divide the population into three groups of type AA, Aa, and aa, respectively. The gene pool carried by this population is assumed to be in proportion p_n of type A in the nth generation. It follows (see Crow and Kimura, 1970) that

$$p_{n+1} = p_n + \frac{p_n(1 - p_n)[(w_{11} - w_{12})p_n + (w_{21} - w_{22})(1 - p_n)]}{w_{11}p_n^2 + 2w_{12}p_n(1 - p_n) + w_{22}(1 - p_n)^2},$$

where w_{11}, w_{12}, and w_{22} are relative fitnesses of the genotypes AA, Aa, and aa, respectively.

If the selective pressures are acting slowly, i.e., if $w_{11} = 1 + \varepsilon\alpha$, $w_{12} = 1$, $w_{22} = 1 + \varepsilon\beta$, where ε is near zero, then

$$p_{n+1} = p_n + \frac{\varepsilon p_n(1 - p_n)[(\alpha - \beta)p_n + \beta]}{1 + O(\varepsilon)}.$$

Regression Analysis

In section 7.1.5, we consider the Munro–Robins algorithm for approximating the root of a function $g(w)$, in the presence of noise. This algorithm corresponds to the following singularly perturbed recurrence relation.

$$w(k + 1) = w(k) - \varepsilon\alpha_k[g(w(k)) + \sigma z_k].$$

Here σ and the α_k, $k = 0, 1, \dots$ are scalars with properties to be specified. We may view $g(w(k)) + \sigma z_k$ as a noisy measurement of $g(w(k))$.

1.2. REVIEW OF THE CLASSICAL LINEAR MULTISTEP THEORY

1.2.1. *Motivation*

A well-known family of numerical methods for dealing with differential equations is the class of *linear multistep methods*. The earliest attacks on stiff problems proceeded with methods from this class. For many problems this is still a good way to proceed. Many further developments for the stiff problem (both historically and in this text) are stimulated by the successes and conspicuous failures of this class concerning the stiff problem. Since moreover, many basic ideas which we make use of such as precision, accuracy, stability and convergence for numerical methods are so well portrayed by the theory of linear multistep methods, we insert a review of this theory at this point.

For details and proofs concerning the theory to be reviewed here, we refer to Henrici, 1962, Isaacson and Keller, 1966, Dahlquist, Bjorck and Anderson, 1974 or to any one of a number of other standard texts. However in Section 4.2, a discussion with the context of the stiff problem is given from which proofs of most of the notions discussed here are directly extractable.

1.2.2. *The Initial Value Problem*

We begin by considering the nonlinear initial value problem

$$\begin{aligned} \dot{x} &= f(t, x), \\ x(a) &= s, \end{aligned} \tag{1.2.1}$$

where x, f and $s \in C_m$ (i.e., are m-tuples of complex numbers). We seek a solution to (1.2.1) on the interval $\mathscr{I}$;

$$\mathscr{I} = \{t \,|\, a \leq t \leq b; \infty < a < b < \infty\}.$$

The class of functions s, for which (1.2.1) is a *well-posed problem* is specified in the following definition.

DEFINITION 1.2.1. f is said to be an L-function (alternatively, $f \in L$) if for all $t \in \mathscr{I}$ and x and $y \in C_m$, there exists a constant L (the Lipschitz constant) such that

$$\| f(t,x) - f(t,y) \| \leq L \| x - y \|.$$

We may now state the following existence and uniqueness theorem for the problem (1.2.1).

THEOREM 1.2.2. *If f is continuous in t for $t \in \mathscr{I}$ and if f is an L-function, the problem (1.2.1) has one and only one solution in $\mathscr{I}$.*

1.2.3. Linear Multistep Operators

The *linear multistep operator* $\mathscr{L}$ is given by

$$\mathscr{L} = \sum_{j=0}^{k} \alpha_j H^j - h \sum_{j=0}^{k} \beta_j H^j \frac{d}{dt}.$$

Here H is the *shift operator*,

$$Hx(t) = x(t+h),$$

and the α_j and β_j are given scalars with $(\alpha_0^2 + \beta_0^2) \cdot \alpha_k \neq 0$. k *is called the number of steps* or the *step number of $\mathscr{L}$*.

The notion of degree of precision is the subject of the following definition.

DEFINITION 1.2.3. $\mathscr{L}$ is said to have *degree of precision p*, if $\mathscr{L}$ annihilates all monomials t^ν, $\nu \leq p$ and p is maximal with respect to this property.

Now let us suppose that $x(t) \in C^\infty$, and let us express $\mathscr{L}x(t)$ in the form of a Taylor series.

$$\mathscr{L}x(t) = \sum_{\nu=0}^{\infty} c_\nu h^\nu x^{(\nu)}(t). \tag{1.2.2}$$

An alternate specification of the degree of precision of $\mathscr{L}$ is given in the following definition.

DEFINITION 1.2.4. $\mathscr{L}$ is said to have degree of precision p if the coefficients α_j and β_j of $\mathscr{L}$ are chosen so that $c_v = 0$, $v = 0, 1, \dots, p$ and p is maximal with respect to this property. Clearly $p \leq 2k$.

1.2.4. Approximate Solutions

To construct an *approximate solution* to (1.2.1), we begin by introducing the *mesh* $t_n = a + nh$, $h > 0$, $n \in J_h \equiv \{0, 1, \dots, n_{max}\}$. J_h is the set of integers n, such that $t_{n+i} \in \mathscr{I}$, $i = 0, 1, \dots, k$.

An approximate solution is a sequence $\{x_n\}$, $n \in J_h$, where x_n is viewed as an approximation to $x(t_n)$, $n \in J_h$. We define an approximate solution by means of $\mathscr{L}$ through the *linear multistep method*,

$$\mathscr{F}(x_n) \equiv \sum_{j=0}^{k} \alpha_j x_{n+j} - h \sum_{j=0}^{k} \beta_j f_{n+j} = 0, \quad n \in J_h. \tag{1.2.3}$$

Here $f_n \equiv f(t_n, x_n)$.

The linear multistep method is said to be *explicit* if $\beta_k = 0$. Otherwise it is *implicit*. Each x_{n+k}, $n \in J_h$ is obtained from (1.2.3) through transposing and solving an equation of the form

$$\alpha_k x_{n+k} - h \beta_k f(t_{n+k}, x_{n+k}) = \text{const.} \tag{1.2.4}$$

In the explicit case, solving this equation requires only division by α_k. Indeed if we normalize $\mathscr{L}$ by dividing it by α_k, even this division may be eliminated. In the implicit case, (1.2.4) represents a system of finite equations which must be solved by a more or less elaborate numerical procedure. (See Section 2.3.)

The linear multistep formula allows the step by step determination of x_n, $n \in J_h$, provided that the values of $x_0, \dots, x_{k-1}$ are known. These so called *starting values* are determined by some independent procedure which may be called the *starting procedure*. As a notation for the starting procedure, we write

$$x_m = S_m(h), \quad m = 0, 1, \dots, k - 1. \tag{1.2.5}$$

Basic properties of the starting procedure are given in the following two definitions.

DEFINITION 1.2.5. The starting procedure is said to be *bounded* if there exists a positive constant M such that $\|S_m(h)\| \leq M$ for all sufficiently small h.

DEFINITION 1.2.6. The starting procedure is said to be *compatible* if

$$\lim_{h \to 0} S_m(h) = s, \qquad m = 0, 1, \ldots, k - 1.$$

Let (see (1.2.1))

$$h_0 = \alpha_k (\beta_k L)^{-1}. \tag{1.2.6}$$

The existence and uniqueness of the numerical procedure which we have just described is the subject of the following theorem.

THEOREM 1.2.7. *A linear multistep formula has one and only one solution* x_n, $n \in J_h$ *for all starting procedures* $S_m(h)$ *if* $0 \le h < h_0$.

1.2.5. Examples of Linear Multistep Methods

The following are some of the well-known linear multistep methods.
(i)   Adams' method:

$$x_{n+k} - x_{n+k-1} - h \sum_{j=0}^{k} \beta_j f_{n+j} = 0.$$

$\beta_k \ne 0$: Adams–Moulton, $k = 1$: Trapezoidal formula.
$\beta_k = 0$: Adams–Bashforth, $k = 1$: Euler's formula.
(ii) Nystrom's method:

$$x_{n+k} - x_{n+k-2} - h \sum_{j=0}^{k-1} \beta_j f_{n+j}.$$

$k = 2$: mid-point formula.
(iii) Method of Newton–Cotes:

$$x_{n+k} - x_n - h \sum_{j=0}^{k} \beta_j f_{n+j} = 0.$$

$k = 2$: Simpson's formula.
(iv) Backward differentiation formula:

$$\sum_{j=0}^{k} \alpha_j x_{n+j} - h \beta_k f_{n+k} = 0.$$

1.2.6. Stability, Consistency and Convergence

Order and *consistency* of a linear multistep method is specified in the following definition.

DEFINITION 1.2.8. A linear multistep method is said to be of order p if

$$\| \mathscr{F}(x(t_n)) \| = O(h^p), \qquad n \in J_h, \tag{1.2.7}$$

where $x(t)$ is any solution of $x' = f(t, x)$ (see (1.2.3)). It is said to be consistent if (1.2.7) is valid with $p \geq 1$.

We now introduce the ρ and σ polynomials.

$$\rho(\omega) = \sum_{j=0}^{k} \alpha_j \omega^j,$$
$$\sigma(\omega) = \sum_{j=0}^{k} \beta_j \omega^j, \tag{1.2.8}$$

and we suppose that $(\rho \,|\, \sigma) = 1$ (i.e., that ρ and σ are *relatively prime*). The following theorem connects these polynomials with the notion of consistency.

THEOREM 1.2.9. *A linear multistep method is consistent if and only if*

$$\mathscr{L}(1) = \rho(1) = 0$$

and

$$\mathscr{L}(t) = h(\rho'(1) - \sigma(1)) = 0.$$

The stability of a linear multistep method is characterized in the following definition.

DEFINITION 1.2.10. A linear multistep formula is said to be *stable* if there is a positive constant M such that

$$\max_{n \in J_h} \| x_n \| \leq M,$$

uniformly in $h, h \in (0, h_0]$ for all bounded starting procedures and for all $f \in L$.

The study of stability makes use of the *root condition* which is given in the following definition.

DEFINITION 1.2.11. A polynomial $\rho(\omega)$ is said to satisfy the root condition if all of its roots lie in the closed unit disc while those on the boundary of the disc are simple. The root condition is sometimes given for the polynomial $\omega^k \rho(\omega^{-1})$. For the latter, the interior and exterior of the unit disc are exchanged; the theory being otherwise identical.

With this definition, we have the following theorem connecting stability and the root condition.

THEOREM 1.2.12. *A linear multistep method is stable if and only if $\rho(\omega)$ obeys the root condition.*

The *global* or *cumulative error* of the linear multistep method is

$$e_n = x_n - x(t_n), \qquad n \in J_h.$$

A *convergent method* is characterized in the following definition.

DEFINITION 1.2.13. A linear multistep method is convergent if for all $f \in L$ and all compatible starting procedures, we have

$$\lim_{h \to 0} \max_{n \in J_h} \|e_n\| = 0.$$

Finally, the main theorem of this subject is the following.

THEOREM 1.2.14. *A consistent linear multistep method is convergent if and only if it is stable.*

Chapter 2

Methods of Absolute Stability

Summary

In this chapter, we discuss the first systematic attacks on the stiff problem. These proceed by imposing strong stability notions onto the traditional linear multistep methods. While this approach has had some success for dealing with the stiff problem, it is perhaps more noteworthy in demonstrating the nature of the limitations of the classical numerical methods; these limitations having stimulated much of the subsequent development of the subject.

We begin with the well-known notion of A-stability in Section 2.1. Then in Section 2.2, we discuss $A(\alpha)$-stability and stiff stability, two of a large number of alternative stability notions associated with stiff problems. Finally in Section 2.3, we consider the problem of solving the equations which are generated by the numerical methods themselves.

2.1. STIFF SYSTEMS AND A-STABILITY

The use of A-stability as a notion for dealing with the stiff problem as well as the theory describing its value and its limitations with respect to the class of linear multistep methods is due largely to G. Dahlquist, 1963. We begin our exposition of these ideas with a motivating discussion.

2.1.1. *Motivation*

Consider the linear system,

$$\dot{x} = Ax, \qquad t \in (0, \bar{t}], \tag{2.1.1}$$

where A is an $m \times m$ constant matrix. Let $\lambda_j, j = 1, \ldots, m$ be the eigenvalues of A. For the following discussion, we take the following definition for characterizing a stiff system.

DEFINITION 2.1.1. The linear system (2.1.1) is said to be *stiff* if

$$\max_{1 \le j \le m} |\lambda_j \bar{t}| \gg 1.$$

Referring to Remark 1.1.2, we see this is not a precisely defined notion. Moreover, this definition obliges us to make the following observation.

REMARK 2.1.2. A system consisting of a single equation may be stiff.

To motivate the first method for dealing with stiff systems, consider the case $m = 2$ with $\lambda_m \ll \lambda_1 < 0$ and with the solution component

$$F(t) = e^{\lambda_1 t} + e^{\lambda_m t}.$$

As t increases from zero there is a *transitory stage* during which $F(t)$ varies extremely rapidly. After a time of the order λ_m^{-1}, the term $e^{\lambda_m t}$ of $F(t)$ becomes negligible, and a new *permanent stage* develops. To determine a numerical approximation to $F(t)$ in the transitory stage, we would use a mesh increment, h_1, such that $|h_1 \lambda_m|$ is small. For the permanent stage, we would like to use a much larger mesh increment h_2 and one such that

$$|\lambda_2 h_2| \ll 1 \ll |\lambda_m h_2|.$$

In this case, the numerical theory described in Section 1.2 is applicable for the term $e^{\lambda_1 t}$. We do not expect the same to be true for the other term. However, if the method is stable no matter how large $|\lambda_m h_2|$ is, we may expect the term $e^{\lambda_m t}$ to remain negligible. This technique calls for methods of an extraordinarily stable character; indeed it calls for methods with a form of absolute stability.

We give three criticisms of this idea.

(i) Getting through the transitory stage requires a number of steps proportional to λ_m^{-1}, and this may not be acceptable. (This number of steps may be considerably reduced by a gradual increase of mesh increment. However, any policy for altering a computation adaptively brings along additional computational cost of its own.)

(ii) If λ_m is large in magnitude because it has a large imaginary part, the so-called transitory stage is itself permanent.

(iii) Absolutely stable methods of simple types are rare. (This will be seen presently.)

For the time being, we exclude eigenvalues with a large imaginary part, and we will return to this type of problem in Section 4.3 and more systematically in Chapter 6.

2.1.2. A-stability

In the following definition, we formalize the well-known notion of absolute stability called *A-stability*.

DEFINITION 2.1.3. A linear multistep method is *A-stable* if all solutions of the difference equation generated by the application of this method to the *test equation* (scalar)

$$\dot{x} = \lambda x, \qquad \lambda \text{ a complex constant,} \qquad (2.1.2)$$

tend to zero as $n \to \infty$ for all λ with $\mathrm{Re}\,\lambda < 0$ and for all fixed $h > 0$.

To determine which linear multistep methods are *A*-stable, we note that when the test equation (2.1.2) is inserted into the linear multistep formula, a linear difference equation results:

$$\sum_{j=0}^{k} (\alpha_j - q\beta_j)y_{n+j} = 0, \qquad q = \lambda h. \qquad (2.1.3)$$

The *characteristic equation* corresponding to (2.1.3) is

$$X(\omega;q) \equiv \rho(\omega) - q\sigma(\omega) = 0. \qquad (2.1.4)$$

(See (1.2.7).)

X defines a k-valued mapping of q into ω. The inverse of this mapping

$$q(\omega) = \rho(\omega)/\sigma(\omega), \qquad (2.1.5)$$

defines a single valued mapping of ω into q.

Having made these observations, we may state the following proposition connecting *A*-stability and the mapping X.

PROPOSITION 2.1.4. *Let* ω_i, $i = 1, \ldots, k$ *be the roots of* $X(\omega;q) = 0$. *Then the following three statements are equivalent.*

(a) *a linear multistep method is A-stable.*
(b) $\mathrm{Re}\,q < 0 \Rightarrow |\omega_i| < 1, \qquad i = 1, \ldots, k.$ $\qquad (2.1.6)$
(c) $|\omega| \geq 1 \Rightarrow \mathrm{Re}\,q(\omega) \geq 0.$

Proof. We forego displaying the proof since it is immediate. □

In the succeeding discussion, we repeatedly refer to the exterior of the unit disc in the ω-plane. So we name this set W, i.e.,

$$W = \{\omega \,|\, |\omega| > 1\}. \qquad (2.1.7)$$

Using Proposition 2.1.4, we may state and prove the following lemma which relates A-stability to the ρ and σ polynomials.

LEMMA 2.1.5. *The linear multistep method* $\sum_{j=0}^{k} (\alpha_j - q\beta_j)x_{n+j} = 0$ *is A-stable only if*
(i) *The roots* σ_i *of* $\sigma(\omega)$ *satisfy* $|\sigma_i| \leq 1$, $i = 1, \ldots, k$.
It is A-stable if an only if
(ii) $\operatorname{Re} \rho(\omega)/\sigma(\omega) \geq 0$, *for all* $\omega \in W$.

Proof. (A) We first show that A-stability implies (i) and (ii).

That A-stability implies (ii) is obvious. We proceed to verify (i). $\rho(\sigma_i) \neq 0$ since $(\rho | \sigma) = 1$. Thus under the mapping of ω into q generated by $X(\omega, q) = 0$, each σ_i is mapped into the north pole of the Riemann q-sphere, the latter being a point on the imaginary axis of that sphere. Similarly, each neighborhood of σ_i is mapped onto a neighborhood of the north pole. Now every neighborhood of the north pole contains values of q such that $\operatorname{Re} q < 0$.

Suppose (i) were not true. Then one of the roots σ_i is such that $|\sigma_i| > 1$. Then there exists a sufficiently small neighborhood of this σ_i contained in W. (See Figure 2.1-1).

Thus, $X = 0$ would have solutions in W for values of q with $\operatorname{Re} q < 0$. This contradicts the A-stability, completing part (A) of this proof.

(B) (ii) implies (2.1.6c) in W. Thus there remains only to verify (2.1.6c) for $|\omega| = 1$. Then let ω_0 be such that $|\omega_0| = 1$ and consider two cases; Case (a) $\sigma(\omega_0) \neq 0$ and Case (b) $\sigma(\omega_0) = 0$.

Case (a): $\sigma(\omega_0) \neq 0$.

In this case, $q(\omega)$ is analytic in a neighborhood of ω_0. Suppose to the

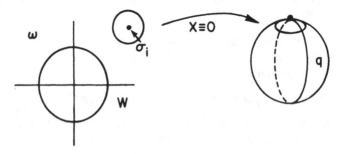

Fig. 2.1-1.

Fig. 2.1-2.

contrary that $\operatorname{Re} q(\omega_0) < 0$. Then a sufficiently small neighborhood of ω_0 will be mapped onto a neighborhood of $q(\omega_0)$, the latter neighborhood being entirely contained in $\operatorname{Re} q < 0$. (See Figure 2.1-2.)

This neighborhood of ω_0 contains points ω of W whose image $q(\omega)$ satisfies $\operatorname{Re} q < 0$. This contradicts (ii) completing the proof of Case (a).

Case (b): $\sigma(\omega_0) = 0$.
In this case, $q(\omega_0)$ is the north pole of the Riemann q-sphere, a point on the imaginary axis. Thus, (2.1.6c) is obviously satisfied. This completes the proof of Case (b) and of the lemma.   □

The following proposition is interesting because it increases the similarity of conditions on $\sigma(\omega_0)$ for A-stability to the root condition (see Definition 1.2.11) imposed on $\rho(\omega)$ for ordinary stability of the linear multistep method.

PROPOSITION 2.1.6. *If a root ω_0 of $\sigma(\omega)$ has magnitude unity and is not a simple root, then the linear multistep method is not A-stable.*
Proof. Let $m \geq 2$ be the multiplicity of the root ω_0. Then $q(\omega) = const(\omega - \omega_0)^{-m}(1 + o(1))$. Thus, the sectors of a neighborhood of ω_0 which are of angle $2\pi/m$ are mapped onto a neighborhood of the north pole of the q-sphere. Since these sectors are at most a half plane, we may choose one which lies entirely in W (except of course for the vertex ω_0 of this sector). (See Figure 2.1-3.) Thus, there exist points of W whose images satisfy $\operatorname{Re} q < 0$. Thus, the corresponding linear multistep method is not A-stable.   □

2.1.3. Examples of A-stable Methods

We now give several examples of A-stable methods.

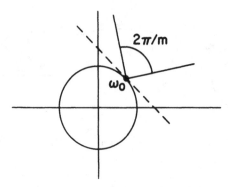

Fig. 2.1-3.

(1) The *trapezoidal formula*:

$$x_{n+1} - x_n - \frac{1}{2}h(f_{n+1} + f_n) = 0.$$

$$\rho(\omega) = \omega - 1, \qquad \sigma(\omega) = \frac{1}{2}(\omega + 1).$$

$$\operatorname{Re} q(\omega) = \frac{|\omega|^2 - 1}{|\omega + 1|^2}.$$

Thus, $\operatorname{Re} q(\omega) > 0$ for $|\omega| > 1$, and the root of σ on $|\omega| = 1$ is simple.

(2) The *backward Euler formula*:

$$x_{n+1} - x_n - hf_{n+1} = 0.$$
$$\rho(\omega) = \omega - 1, \qquad \sigma(\omega) = \omega.$$
$$\operatorname{Re} q(\omega) = \frac{|\omega|^2 - \operatorname{Re} \omega}{|\omega|^2} > 0, \qquad |\omega| > 1.$$

(3) $$x_{n+k} - x_n - \frac{1}{2}hk(f_{n+k} + f_n) = 0.$$

$$\rho(\omega) = \omega^k - 1, \qquad \sigma(\omega) = \frac{1}{2}k(\omega^k + 1).$$

The roots of $\sigma(\omega)$ are the kth roots of unity.

$$\operatorname{Re} q(\omega) = \frac{1}{2}k\frac{|\omega|^{2k} - 1}{|\omega^k + 1|^2} > 0, \qquad |\omega| > 1.$$

Note further that $\rho(1) = 0$, $\rho'(1) = \sigma(1) = k$, implying the consistency of this method (see Theorem 1.2.8). This example shows *the existence of linear multistep methods which are consistent and A-stable for any k* (i.e., for any number of steps).

2.1.4. Properties of A-stable Methods

Achieving A-stability is costly in terms of the restrictions this property imposes on the class of linear multistep methods. The first restriction is the loss of explicit schemes which requires a greater amount of computation in each step of the method. (See (1.2.4)f.) This restriction is characterized by the following theorem.

THEOREM 2.1.7. *An explicit multistep method cannot be A-stable.*

Proof. Assume to the contrary that the method is both explicit and A-stable. Then $\beta_k = 0$ and $q(\omega) = \rho(\omega)/\sigma(\omega)$ has a pole at the point, ω_0, at infinity on the ω-sphere. But ω_0 as well as a neighborhood of ω_0 lie in W. The image of such a neighborhood under the mapping $q = q(\omega)$ is a neighborhood of the point, q_0, at infinity on the q-sphere. Such a neighborhood contains points for which Re $q < 0$. This contradicts (2.1.6c) completing the proof of the theorem.   □

If a linear multistep method is of order p (see Definition 1.2.8), we have from (1.2.3) that

$$\mathscr{L}(x(t)) = c_{p+1} h^{p+1} x^{(p+1)}(t)(1 + O(h)).$$

If $p \geq 1$, $\rho(1) = 0$, and since $(\rho|\sigma) = 1$, then $\sigma(1) \neq 0$. Now consider the following definition and remark which introduce the so-called error constant c^*, which serves as a measure of quality of linear multistep methods of the same order.

DEFINITION 2.1.8. $c^* = -c_{p+1}/\sigma(1)$ is called the *error constant* of a linear multistep method of order $p > 1$.

REMARK 2.1.9. $c^* = \lim_{\omega \to 1} \left[\log \omega - \rho(\omega)/\sigma(\omega) \right]/(\omega - 1)^{p+1}.$ (2.1.8)

The following theorem characterizes the key restriction on A-stable methods.

THEOREM 2.1.10. *The order p of an A-stable linear multistep method cannot exceed 2. The trapezoidal formula is the A-stable method or order 2 which supplies the smallest error constant, $c^* = 1/12$.*

Proof. The proof begins with a side calculation.

Let $z = (\omega + 1)/(\omega - 1)$, (the well known 1-1 Moebious transformation, carrying $\omega = 1$ into the point z at infinity). Let the transformation Γ be defined by

$$\Gamma f(\omega) = 2^{-k/2}(\omega - 1)^k f\left(\frac{\omega + 1}{\omega - 1}\right),$$

and let

$$r(z) = \Gamma\rho(\omega), \qquad s(z) = \Gamma\sigma(\omega).$$

Now apply Γ to (2.1.8). We get

$$\log\frac{z+1}{z-1} - \frac{r(z)}{s(z)} = c^*\left(\frac{2}{z}\right)^{p+1}(1 + o(1)), \qquad z \to \infty.$$

Since

$$\log\frac{z+1}{z-1} = 2z^{-1} + \frac{2}{3}z^{-3} + O(z^{-4}),$$

this becomes

$$\frac{r(z)}{s(z)} = 2z^{-1} + \left(\frac{2}{3} - 8c'\right)z^{-3} + O(z^{-4}), \tag{2.1.9}$$

where

$$c' = \begin{cases} c^*, & p = 2, \\ 0, & p \ge 3. \end{cases}$$

Thus we may note that the coefficient of z^{-3} in (2.1.9) is strictly positive if $p \ge 3$.

Next we translate the conditions (i) and (ii) of Lemma 2.1.5. By using properties of the Moebious transformation, we see that this lemma shows that the A-stability of a linear multistep method implies the following conditions (i) and (ii).

(i) The roots s_i of $s(z)$ satisfy Re $s_i \le 0, i = 1, \ldots, k$.

(ii) Re $(r(z)/s(z)) \ge 0$ for all z in Re $z \ge 0$.

$$\tag{2.1.10}$$

Next we make use of the following variant of the Riesz-Herglotz theorem (see Achiezer and Glassman, 1959, p. 152):

THEOREM 2.1.11. *An analytic function $\phi(z)$ which satisfies the following conditions:*

(a) $\sup\limits_{0<x<\infty} |x\phi(x)| < \infty,$

(b) $\phi(z)$ *regular in* Re $z > 0,$

(c) Re $\phi(z) \geq 0$ *in* Re $z > 0,$

may be represented as follows.

$$\phi(z) = \int_{-\infty}^{\infty} \frac{d\omega(t)}{z - it},$$

where $\omega(t)$ is a bounded nondecreasing function.

We will show that $z\phi(z) = zr(z)/s(z)$ is bounded for all $x \in [0, \infty]$. We note first that (2.1.9) implies that $xr(x)/s(x)$ is bounded as $x \to \infty$. By hypothesis the linear multistep method is A-stable. Then from Proposition 2.1.6, $\sigma(\omega)$ has a zero of order at most unity at $\omega = -1$. The same then is true for $s(z)$ at $z = 0$. Then $xr(x)/s(x)$ is bounded at $x = 0$. Using (2.1.10)(i), we may conclude that $xr(x)/s(x)$ is bounded for all x on the positive real axis. Thus, $z\phi(z)$ is indeed bounded as claimed. Re $\phi(z) \geq 0$ in the half plane Re $z > 0$. Thus the hypotheses (a), (b) and (c) of the cited Theorem 2.1.11 are verified for $x > 0$, and we have

$$x\frac{r(x)}{s(x)} = \int_{-\infty}^{\infty} \frac{x}{x - it}\, d\omega(t) = \int_{-\infty}^{\infty} \frac{x^2}{x^2 + t^2}\, d\omega(t).$$

Since

$$\frac{d}{dx} \frac{x^2}{x^2 + t^2} = \frac{2xt^2}{(x^2 + t^2)^2} \geq 0 \text{ for } x \geq 0,$$

we may conclude from this representation that

$$\frac{d}{dx}\left[x\frac{r(x)}{s(x)} \right] \geq 0. \tag{2.1.11}$$

Next from (2.1.9) we may conclude that

$$\frac{d}{dx}\left[x\frac{r(x)}{s(x)} \right] = -2(2/3 - 8c')x^{-3}(1 + O(1)), \qquad x \to \infty. \tag{2.1.12}$$

Comparing (2.1.11) and (2.1.12), we deduce that

$$2/3 - 8c' \leq 0 \tag{2.1.13}$$

If $p \geq 3$, $c' = 0$ and (2.1.13) is impossible. This demonstrates the first assertion of the theorem.

If $p = 2$, then $2/3 - 8c' \leq 0$ or $c^* \geq 1/12$. For the trapezoidal formula,

$$\rho(\omega) = \omega - 1, \sigma(\omega) = \frac{1}{2}(\omega + 1), r(z) = \sqrt{2}, s(z) = z/\sqrt{2},$$

so that

$$\frac{r(z)}{s(z)} = \frac{2}{z}.$$

Comparing this with (2.1.9), we deduce that $2/3 - 8c^* = 0$ or $c^* = 1/12$. This demonstrates the second assertion of the theorem and completes its proof.   □

2.1.5. A Sufficient Condition for A-stability

Condition (ii) of Lemma 2.1.5 requires the verification of a property of $q(\omega)$ for all ω in W (see (2.1.7)). A less stringent requirement furnishes the sufficient condition for A-stability characterized by the following theorem.

THEOREM 2.1.12. If (i) the roots σ_i of $\sigma(\omega)$ satisfy $|\sigma_i| < 1, i = 1, \ldots, k$ and (ii) $u(\omega) \equiv \text{Re } q(\omega) \geq 0$ on the unit circle, then the linear multistep is A-stable.
Proof. (i) implies that $q(\omega)$ is analytic in $\overline{W}$ and in particular at $\omega = \infty$. Then $u(\omega)$ is harmonic in $\overline{W}$, and from the minimum principle

$$u(\omega) \geq \min_{|\omega| = 1} u(\omega),$$

for all $\omega \in \overline{W}$. Then (ii) implies that $u(\omega) \geq 0$ for all $\omega \in \overline{W}$. Then (2.1.6c) implies that the method is A-stable completing the proof of the theorem.   □•

2.1.6. Applications

As an application of Theorem 2.1.12 consider the formula

$$x_{n+1} - x_n - h[(1 - a)\dot{x}_{n+1} + a\dot{x}_n] = 0, \tag{2.1.14}$$

for which $p \geq 1$ for all real values of the parameter a. For $a = 1, 1/2, 0$, respectively, this formula becomes the Euler formula, the trapezoidal formula and the backward Euler formula, respectively. In any case, we have

$$\sigma(\omega) = (1 - a)\omega + a.$$

The root $\sigma_1 = -a(1-a)^{-1}$ of $\sigma(\omega)$ is less than unity in magnitude if and only if $a < 1/2$. A calculation shows that

$$u(e^{i\theta}) = |\sigma(e^{i\theta})|^{-2} P(e^{i\theta}),$$

where

$$P(e^{i\theta}) = (1 - 2a)(1 - \cos\theta).$$

$P(e^{i\theta}) \geq 0$ if and only if $a \leq 1/2$. Thus (2.1.14) is A-stable if $a < 1/2$.

Note that the trapezoidal formula (which is A-stable) fails to satisfy the sufficient condition of Theorem 2.1.12. A second application of Theorem 2.1.12 is furnished by the following formula.

$$(-1 - a + b)x_n + 2(a - b)x_{n+1} + (1 - a + b)x_{n+2}$$
$$- h[a\dot{x}_n + (2 - a - b)\dot{x}_{n+1} + b\dot{x}_{n+2}] = 0.$$

For this formula, $p \geq 2$ for all real values of the parameters a and b. One may show that for this formula, hypotheses (i) and (ii) of Theorem 2.1.12 are equivalent to the following inequalities.

$$b - a > 0,$$
$$-1 + a + b > 0,$$

2.2. NOTIONS OF DIMINISHED ABSOLUTE STABILITY

The family of linear multistep methods is so desirable because of its simple form for both computation and analysis that the limitations imposed on this family by A-stability made a great impact. To attempt to save the family for the solution of stiff differential equations, a sequence of weakened forms of absolute stability were invented in order. We will examine two of these. First we will consider the notation of $A(\alpha)$-stability, and following that, we will review the so-called stiffly stable methods.

2.2.1. $A(\alpha)$-stability

Examination of the failure of linear multistep methods to be A-stable shows in many cases that failure occurs for values of λ in the test equation which are nearly purely imaginary. It is then a simple step to abandon such values of λ (i.e., highly oscillatory solutions) and to seek the analogue of A-stability corresponding to a subset of the left half complex plane which in particular excludes the imaginary axis. The most straightforward approach to doing this is to replace the left half plane by a cone S_α, with

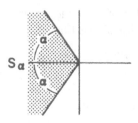

Fig. 2.2-1.

vertex at the origin centered on the negative half axis and of half angle α. (See Figure 2.2–1.)

With such a cone in mind, consider the following definition.

DEFINITION 2.2.1. A linear multistep method is $A(\alpha)$-*stable*, $0 < \alpha < \pi/2$, if all solutions of the difference equation arising through the application of this method to the test equation (see Definition 2.1.3) tend to zero as $n \to \infty$ for each fixed mesh increment $h > 0$ and for all $\lambda \neq 0$, where

$$q \equiv \lambda h \in S_\alpha \equiv \{q \mid |\arg(-q)| < \alpha, \quad q \neq 0\}.$$

We make the following observations concerning $A(\alpha)$-stability.

REMARK 2.2.2. Let ω_i, $i = 1, \dots, k$ be the roots of the characteristic equation $X = 0$, corresponding to the difference equation arising from the application of the test equation to the linear multistep method. Then the corresponding linear multistep method is $A(\alpha)$-stable if $q \in S_\alpha$ implies that the $|\omega_i| < 1, i = 1, \dots, k$.

REMARK 2.2.3. (a) $A(\alpha)$-stability $\Rightarrow A(\beta)$-stability for $0 < \beta < \alpha$. (b) A-stability is equivalent to $A(\pi/2)$-stability.

The case $\alpha = 0$ is described in the following definition.

DEFINITION 2.2.4. A linear multistep method is $A(0)$-*stable* if it is $A(\alpha)$-stable for all sufficiently small $\alpha > 0$.

The following lemma which employs the functions $r(z)$ and $s(z)$ introduced in the proof of Theorem 2.1.10, is the analogue of Lemma 2.1.5.

LEMMA 2.2.5. *The linear multistep method* $\sum\limits_{j=0}^{k} (\alpha_j - q\beta_j)x_{n+j}$ *is* $A(\alpha)$-*stable,* $\alpha > 0$, *if*

(i) *The roots* s_i *of* $s(z)$ *satisfy* Re $s_i \leq 0, i = 1, \ldots, k$.
It is $A(\alpha)$-*stable if and only if*
(ii) $r(z)/s(z)$ *is in the compliment of* S_α *for all* z *with* Re $z > 0$.

For the case of $A(0)$-stability, we have the following necessary condition.

LEMMA 2.2.6. *If a linear multistep method is* $A(0)$-*stable then* $\{\alpha_\nu \geq 0$ *or* $\alpha_\nu \leq 0\}$ *and* $\{\beta_\nu \geq 0$ *or* $\beta_\nu \leq 0\}$, $\nu = 1, \ldots, k$.

We forego developing the proofs of Lemmas 2.2.5 and 2.2.6 since the proofs are generally analogous to the proofs in Section 2.1. (Alternatively, see Widlund, 1967.)

2.2.2. Properties of $A(\alpha)$-stable Methods

As usual we will suppose that $(\rho \,|\, \sigma) = 1$ and that $p \geq 1$ (so that the methods are consistent).

The following theorem shows that abandonment of wedges adjacent to the imaginary axis is not sufficient to recover the explicit methods which were already lost to A-stability.

THEOREM 2.2.7. *An explicit linear multistep method cannot be* $A(0)$-*stable.*

On the other hand the restriction $p \leq 2$ on the order of methods is weakened, at least somewhat, as the following two theorems show.

THEOREM 2.2.8. *The trapezoidal formula is the only* $A(0)$-*stable linear multistep method with* $p \geq k + 1$.

THEOREM 2.2.9. *There exist* $A(\alpha)$-*stable linear multistep methods,* $0 \leq \alpha < \pi/2$, *for* $k = p = 3$ *and* $k = p = 4$.

The proofs of these three theorems are not given since they are analogous to proofs in Section 2.1. (Alternatively, see Widlund, 1967.)

2.2.3. Stiff Stability

The class of methods which are stiffly stable is uncovered by an abandonment of the imaginary axis for the domain of stability. For these methods,

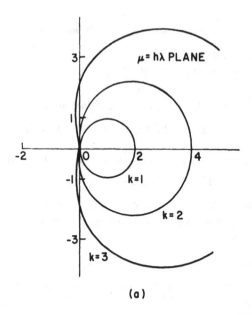

(a)

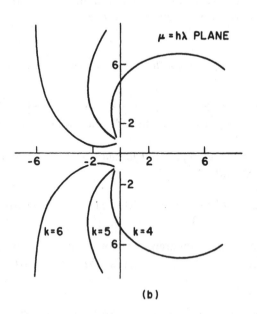

(b)

Fig. 2.2-2. Regions of stability of some stiffly stable methods.

the left half plane of A-stability is replaced by another left half plane contained in the former. Such a method will certainly leave quiescent a component of the form $e^{\lambda_m t}$ (see Section 2.2.1). On the other hand, for a non-stiff mode $e^{\lambda_1 t}$, the method must provide accuracy in a neighborhood S of the origin into which λ_1 may be scaled by multiplication by h, h reasonably small.

With these tactics in mind, consider the following definition.

DEFINITION 2.2.10. A consistent method is *stiffly stable* if the following two conditions hold.

(i) For some constant $D < 0$, all solutions of the difference equation generated by the application of this method to the test equation tend to zero as $n \to \infty$ for all λ with Re $\lambda < D$ and for all fixed $h > 0$.

(ii) There is an open set S whose closure contains the origin such that the method is stable for $h\lambda \in S$.

Among the methods which are stiffly stable are the so-called backward differentiation formulas (see Section 5.1.4 of Henrici, 1962). Low order formulas of this type provide the basis of the well known *Gear's package* for solving stiff differential equations. (See Hindmarsh, 1974.)

The first three backward differentiation formulas are:

$$k = 1 \qquad y_{n+1} - y_n - h\dot{y}_{n+1} = 0,$$

$$k = 2 \qquad y_{n+2} - \frac{4}{3}y_{n+1} + \frac{1}{3}y_n - \frac{2}{3}\dot{y}_{n+2} = 0, \qquad\qquad (2.2.1)$$

$$k = 3 \qquad y_{n+3} - \frac{18}{11}y_{n+2} + \frac{9}{11}y_{n+1} - \frac{2}{11}y_n - \frac{6}{11}h\dot{y}_{n+3} = 0.$$

In Figure 2.2-2a, we plot the regions of stability for these three methods. The methods are stable outside of the curves indicated. In Figure 2.2-2b, regions of stability are plotted for some stiffly stable methods of orders 4, 5 and 6 (see Gear, 1971).

2.3. SOLUTION OF THE ASSOCIATED EQUATIONS

2.3.1. *The Problem*

Applying A-stable, $A(\alpha)$-stable or stiffly stable methods (of the backward differentiation formula type) to the stiff initial value problem forces the linear multistep methods to be implicit (see Theorem 2.1.7, Theorem

2.2.7 and equation (2.2.1)). Thus at each mesh point t_n, we are confronted with the problem of solving a system of equations of the form

$$x - h\beta f(x) = b \qquad\qquad (2.3.1)$$

(compare (1.2.4)), where β is a constant and b is a constant vector. Even in the linear case

$$(I - h\beta A)x = b, \qquad\qquad (2.3.2)$$

the solution of such systems is difficult. This is so because the stiffness of the original system of differential equations translates into ill-conditioning for the system of finite equations. In (2.3.2) and hereafter (and unless otherwise specified), I denotes the m-dimensional identity matrix.

Convergence of an iterative method corresponding to (2.3.2) is equivalent to a condition on the spectral radius of $I - h\beta A$. A typical such condition is that this spectral radius be less than unity, i.e.,

$$0 < h\beta\lambda < 2, \qquad\qquad (2.3.3)$$

for each eigenvalue $\lambda \in \sigma(A)$, the spectrum of A. Of course in the stiff case, this restriction on h is impractical and unacceptable.

A direct method of solution avoids this particular difficulty, but for the *condition number* $\mu(A) = \|A\|/\|A^{-1}\|$, we have

$$\mu(A) = |\lambda_m \lambda_1^{-1}|$$

approximately, where λ_1 and λ_m are the eigenvalues of A of smallest and largest magnitude, respectively. Apart from the possibly large amount of computation required of a direct method of solution, this poor conditioning of the problem makes the numerical solution process unstable and unreliable.

In practice, Newton's method or a variant is used to solve (2.3.1), i.e., an iteration scheme of the form

$$x_{n+1} = x_n + (I - h\beta f_x(x_n))^{-1}(x_n - h\beta f(x_n)) \qquad\qquad (2.3.4)$$

is employed.

It is the rapid (quadratic-)convergence of Newton's method which seems to be the reason why this approach is more successful than the previous ones. Of course to exploit the rapid convergence of Newton's method, a good starting value for the recurrence must be supplied. In the transient state (see the discussion following Remark 2.1.2), this is not so easy to come by, but in the permanent state, such a starting value is easily supplied by extrapolation from the numerical approximation (to the differential equations) at several preceding mesh points. The quality of this starting value is a critical ingredient in this approach to solving (2.3.1).

2.3.2. Conjugate Gradients and Dichotomy

The nonlinear system (2.3.1) may be linearized locally (as in Newton's method (2.3.4)). In this section, we describe several algorithms for solving such a linear system which employ the *method of conjugate gradients*. Some of these algorithms employ a prediction process as well. The algorithms do not appear to be adversely affected by the difficulties which we have just described. Indeed, they seem to be suited to the problem (2.3.2) corresponding to the stiff differential equation.

Conjugate Gradient Algorithm

Let B be a symmetric and positive definite $m \times m$ matrix. Corresponding to the m-dimensional system

$$Bx = b,$$

whose solution we denote by x^*, the conjugate gradient algorithm is defined recursively as follows.

Given x_i, r_i and p_i, compute

$$a_i = \frac{(r_i, r_i)}{(Bp_i, p_i)},$$
$$x_{i+1} = x_i + a_i p_i,$$
$$r_{i+1} = r_i - a_i Bp_i, \tag{2.3.5}$$
$$p_{i+1} = r_{i+1} + b_i p_i,$$

where

$$b_i = \frac{(r_{i+1}, r_{i+1})}{(r_i, r_i)}, \quad i = 1, 2, \dots.$$

The initial values of this recursion are taken to be

$$x_0 = 0, \quad r_0 = p_0 = b.$$

The conjugate gradient algorithm is an iterative scheme which converges in principle in m (the dimension) or fewer steps. Thus in fact, it is an elimination method as well, and as such, its convergence is not determined by a condition of the type (2.3.3) (necessary and sufficient for the convergence of linear iterations).

We now make a *dichotomized characterization* of the spectrum $\sigma(A)$ of A. We suppose that $\sigma(A) = \sigma_L(A) \cup \sigma_S(A)$, and that there exist constants c and C such that

$$0 < c \le 1 \le C,$$

and such that

$$|\lambda| \ge C, \quad \lambda \in \sigma_L(A),$$
$$|\lambda| \le c, \quad \lambda \in \sigma_S(A).$$

Let L and S denote the invariant subspaces of R^m which correspond to eigenvalues in σ_L and in $\sigma_S(A)$, respectively.

When A is symmetric and positive, the conjugate gradient algorithm furnishes favorable properties for systems whose spectrum tends to cluster. Indeed, suppose that the cardinality of $\sigma_L(A)$, $|\sigma_L(A)| = l$. Then $\sigma(A)$ may be viewed as composed of $l+1$ clusters. One cluster is $\sigma_S(A)$ itself while each of the eigenvalues in $\sigma_L(A)$ comprises a separate cluster. In fact if the eigenvalues in $\sigma_L(A)$ tend to cluster among themselves, the situation with respect to the conjugate algorithm will be even further improved.

These claims concerning the conjugate gradient algorithm derive from the following error estimate for s-th iterate x_s, furnished by that algorithm.

$$\| x_s - x^* \| \le \min\,(B^{-1}R(B)b, R(B)b), \tag{2.3.6}$$

where the minimum is taken over all polynomials $R(z)$ of degree s or less and which are normalized by $R(0) = 1$ (see Luenberger, 1965 or Miranker, 1972). Thus the value of $(B^{-1}R(B)b, R(B)b)$, for any such polynomial $R(z)$, provides an upper bound for $\| x_s - x^* \|$.

The computational problem derived in Section 2.3.1, and to which we will apply the conjugate gradient algorithm, is the solution at each mesh point t_n of an equation of the following form.

$$Bx \equiv (I - h\beta A)x = b. \tag{2.3.7}$$

Algorithms
We propose three algorithms for treating (2.3.7).

Alg_0 : Apply the method of conjugate gradients directly to (2.3.7).

Alg_1 : Make a prediction x^1 of the solution of (2.3.7), where

$$x^1 = b + h\beta A b.$$

For the residual of this prediction, $\rho^1 = b - (I - h\beta A)x^1$, we have

$$\rho^1 = h^2\beta^2 A^2 b. \tag{2.3.8}$$

Clearly $\rho^1|_S = O(h^2 c^2 \beta^2)$, but $\rho^1|_L \geq O(h^2 C^2)$. Thus the prediction, while good on S, is very poor on L.

Writing the solution x^* of (2.3.7) as

$$x^* = x^1 + v^1, \tag{2.3.9}$$

we have

$$(I - h\beta A)v^1 = \rho^1. \tag{2.3.10}$$

The algorithm consists of applying the method of conjugate gradients to (2.3.10) to compute v^1, and then to determine x^* from (2.3.9).

Alg_p : This algorithm, which is an extension of Alg_1, begins with the prediction

$$x^p = (I + h\beta A + \ldots + h^p\beta^p A^p)b$$

of the solution to (2.3.7). The corresponding residual ρ^p is

$$\rho^p = (h\beta A)^{p+1} b. \tag{2.3.11}$$

Writing the solution x^* of (2.3.7) as

$$x^* = x^p + v^p, \tag{2.3.12}$$

we have that

$$(I - h\beta A)v^p = \rho^p. \tag{2.3.13}$$

The algorithm consists of applying the method of conjugate gradients to (2.3.13) to compute v^p, and then to determine x^* from (2.3.12).

If $v^{p,(s)}$ is the vector produced by s steps of the conjugate gradient algorithm applied to (2.2.13), and setting $x^{(s)} = x^p + v^{p,(s)}$, notice that

$$x^* - x^{(s)} = x^p + v^p - (x^p + v^{p,(s)}) = v^p - v^{p,(s)}, \quad p = 1, 2, \ldots. \tag{2.3.14}$$

To determine the effectiveness of these algorithms, we employ (2.3.6).

Recalling that $|\sigma_L(A)| = l$, set

$$R(\mu) = \prod_{i=1}^{l} (1 - \mu/\mu_i), \quad \mu_i = 1 - h\beta\lambda_i. \tag{2.3.15}$$

Then we will see that for an approximation produced by l conjugate gradient steps, the algorithms produce the following results, respectively.

$$
\begin{align}
&(0) \quad \| x^{(l)} - x^* \| = O(1) \\
&(1) \quad \| v^{1,(l)} - v^1 \| = O(h^2 c^2 \beta^2) \tag{2.3.16} \\
&(p) \quad \| v^{p,(l)} - v^p \| = O(h^{p+1} c^{p+1} \beta^{p+1}).
\end{align}
$$

We will verify this estimate in the case (2.3.16)1). Since the μ_i are the eigenvalues of B (see (2.3.7) and (2.3.15)),

$$(B^{-1} R(B)\rho^1, R(B)\rho^1) = \sum_{i=1}^{m} \frac{R^2(\mu_i)}{\mu_i} |\rho_i^1|^2,$$

where ρ_i^1 is the component of ρ^1 in the eigenspace corresponding to μ_i. (There is an implicit assumption of simplicity of the eigenvalues made here for reasons of convenience and which is easily avoided.) Since $R^2(\mu_i) = 0$ for $\mu_i \in \sigma_L(B)$, we get

$$
\begin{align}
\| x^* - x^{(l)} \| &\leq \sum_{\{i | \mu_i \in \sigma_s (B)\}} \frac{R^2(\mu_i)}{\mu_i} |\rho_i^1|^2 \\
&\leq O \| \rho^1 |_s \| = O(h^2 c^2 \beta^2), \tag{2.3.17}
\end{align}
$$

by appealing to (2.3.14) for the left member here and to (2.3.8) for the right.

If the eigenvalues in $\sigma_L(B)$ themselves fall into d different clusters $K_j, j = 1, \ldots, d$, the polynomial (2.3.15) may be replaced by

$$R_d(\mu) = \prod_{j=1}^{d} (1 - \mu/\gamma_j),$$

where γ_i is the center of the ith such cluster. Correspondingly, we execute $d < l$ steps of the conjugate gradient algorithm. The error estimate is degraded since $R_d(\mu_i) = 0$ is replaced by

$$\sum_{\mu \in K_i} \prod_{j=1}^{d} (1 - \mu/\gamma_j),$$

a quantity which depends on the diameters of the clusters $K_i, i = 1, \ldots, d$.

The error estimate (2.3.17) shows the value of a prediction. On the subspace S, the prediction annihilates the error in the solution we seek to the accuracy $O(h^{p+1})$ of the numerical method (see (2.3.11)). The conjugate gradient algorithm does not disturb this accomplishment, but proceeds to annihilate the remaining error which is in L, by use of a number of steps not exceeding the cardinality l of $\sigma_L(B)$. If we proceed further with say q conjugate gradient steps, the error bound will be improved by a factor $h\beta c$ per step. This follows since such steps consist principally of solving an equation in S. Since in S, the eigenvalues of $I - \beta hA$ cluster around unity, (i.e., are of the form $1 - h\beta\lambda$; see (2.3.10)), this claim is verified by employing the previous argument and the polynomial

$$R(\mu) = \prod_{i=1}^{l} (1 - \mu/\mu_i)(1 - \mu)^q.$$

Work Needed for the Algorithms

In Table 2.3-1, we list the approximate number of conjugate gradient steps (NCG) required to achieve an error bound of the size $O((hc\beta)^{p+1})$ for each of the three algorithms $Alg_j, j = 0, 1, p$.

(2.3.5) shows that each conjugate gradient step requires a single matrix multiplication, say M units of work and two inner products, say P units of work each. Preprocessing for the prediction step shows that it requires a single matrix multiplication, M units of work. We neglect the minor ancillary arithmetic operations in our count. Since an inner product is $1/m$ times the work of a matrix product, we ignore P compared to M as well. We tabulate the work needed to produce an error bound of the size $(O(hc\beta))^{p+1}$ for the algorithms in Table 2.3-1. In that table, we use the abbreviation

$$W = (l + p + 1)M.$$

TABLE 2.3-1

Alg	NCG	Work
0	$l + p + 1$	W
1	$l + p - 1$	$W - M$
p	l	$W - pM$

A Scaling Question

The prediction step amplifies by factors of the size $O(h^{p+1}\lambda_i^{p+1})$. While these factors are small for the $\lambda_i \in S$, they are enormous for the $\lambda_i \in L$. A large prediction gives a large residual, and in turn, a large correction which must combine (see (2.3.4) or (2.3.12)) to produce a moderate result. This is a well known computational situation to be avoided. The efficacy of prediction as shown in Table 2.3-1 can, however, be redeemed by exploiting a dichotomous characterization of the solution of the differential equation (2.1.1).

As t increases from zero, the solution of the differential equation decays exponentially with the time constants which act algebraically, as we have observed, in the prediction process. Correspondingly as t_n increases as we proceed along the mesh, the global numerical error decays geometrically on L. (See Theorem 2.1, Miranker and Chern, 1980, where this result is demonstrated for the backward differentiation formulas (see (2.2.1)) of any step number k.) Thus the numerical solution will decay likewise on L. Thus at the mesh point t_n, the prediction process, amplifying by factors of the form $O(h^p\lambda^p)$, will deliver a residual of the size

$$O(h^{p+1}\lambda^{p+1})O\left(\frac{1}{1+\beta hC}\right)^{n/k}$$

on L (i.e., for $\lambda \in \sigma_L(A)$). Thus if n is large enough compared to p, this residual is not at all large and the scaling problem is vacuous. Indeed we may see in Table 2.3-2, which shows the result of computational experiments, that $n = 1$ is too small but $n \geq 2$ large enough to control the scaling problem. In particular, notice in this table that the number of conjugate gradient steps (NCG) needed to achieve a specified level of accuracy drops in passing from $n = 1$ to $n = 2$. The degree of improvement is proportional to the degree of prediction, as we expect. These computational experiments are described in the next section.

2.3.3. *Computational Experiments*

In this section, we present the results of computational experiments with Alg_0, Alg_1 and Alg_2. As numerical methods, we employ the backward differentiation formulas for $k = 1, 2$ (see (2.2.1)). We use the two algorithms Alg_0 and Alg_1 in the case $k = 1$ and the three algorithms Alg_0, Alg_1 and Alg_2 in the case $k = 2$.

We take $m = 10$, and for the spectrum, we take $\sigma(-A) = \{10^5, 10^5,$

$10^4, 10^4, 10^3, 1, 10^{-1}, 10^{-2}, 10^{-3}, 10^{-4}\}$. Note that this spectrum has three large clusters, *i.e.*, $l = 3$. The mesh increment is chosen to be $h = 0.1$.

To provide a definite and a symmetric A, and one for which the exact solution of the differential equation could be found, we select a random

TABLE 2.3-2

$k = 1$

n	NCG_0	NCG_1	$TA_0 \times 10^5$	$TA_1 \times 10^5$	TE_0	TE_1
1	7	5	0.244	2.0	0.0277	0.0277
2	6	4	0.487	3.96	0.0307	0.0307
3	6	4	0.729	5.88	0.0418	0.0418
4	6	4	0.97	7.76	0.0505	0.0505
5	6	4	1.21	9.6	0.0573	0.0573

$k = 2$

n	NCG_0	NCG_1	NCG_2	$TA_0 \times 10^3$	$TA_1 \times 10^3$	$TA_2 \times 10^3$	$TE_0 \times 10^3$	$TE_1 \times 10^3$	$TE_2 \times 10^3$
1	5	5	6	0	0	0	0	0	0
2	5	4	3	0	0	0	0	0	0
3	5	4	3	0.896	0.00601	0.056	11.0	11.0	10.9
4	5	4	3	2.07	0.0139	0.839	2.7	1.59	0.775
5	5	4	3	3.32	0.0223	1.83	4.2	2.34	0.523
6	5	4	3	4.57	0.0308	2.83	5.0	2.98	0.265
7	5	4	3	5.82	0.0392	3.07	8.3	3.48	0.0767
8	5	4	3	7.04	0.0474	4.06	9.5	3.87	0.212
9	5	4	3	8.24	0.0555	4.52	1.1	4.15	0.377
10	5	4	3	9.41	0.0634	4.85	5.13	4.34	0.413

$m \times m$ matrix $\bar{S}$. Then $\bar{S}$ is replaced by S, the latter obtained by ortho-normalizing the columns of $\bar{S}$. Finally

$$A = \Lambda S^T,$$

where Λ is the diagonal matrix whose diagonal entries are those of $\sigma(A)$ taken in order. The initial vector is chosen to be

$$x(0) = (1, 2, 3, \ldots, 10)^T.$$

Starting values for the case $k = 2$ are determined by an application of the algorithm for the case $k = 1$.

We call TE the Euclidean norm of the global error, and we call TA the Euclidean norm of the algebraic error (i.e., the error between the exact solution of (2.3.7) and the solution produced by these methods). For purposes of comparison, the exact solution of the differential equation and of (2.3.7), where needed, is produced by some elimination method to a least ten figures.

We call NCG_i the number of conjugate gradient steps needed to produce convergence of $Alg_i = 0, 1, 2$ to a prescribed tolerance, (*viz* h^p).

Discussion
Results of the experiments are displayed in Table 2.3-2. The stopping tolerance is $h^2 = 0.01$. Thus each algorithm uses as many conjugate gradient steps as required to reduce the algebraic error (TA) to 0.01. Examining the row $n = 10$ in the case $k = 2$ of the table, we see that the actual algebraic errors (TA) achieved are 9.41×10^{-3}, 0.0634×10^{-3} and 4.85×10^{-3}, respectively. Such wide variations are characteristic in the table and point out the sensitivity of the algorithms to a single conjugate gradient step. Thus while the value 0.0634×10^{-3} referred to and which is produced by $NCG_1 = 4$ conjugate gradient steps is much smaller than necessary, the fact is that three such steps fail to achieve the tolerance of 0.01 for Alg_1. The corresponding global errors (TE) are less widely varied. In fact the global errors tend to improve with passage from $Alg_0 \rightarrow Alg_1 \rightarrow Alg_2$. Thus the more elaborate algorithm gives a generally better result with less work (lower NCG).

Since $l = 3$ and $p = 2$ in the experiments, Table 2.3-1 predicts that $NCG_2 = 3$, $NCG_1 = 4$ and $NCG_0 = 6$. These actual values are 3, 4 and 5 respectively, in quite good agreement.

Chapter 3

Nonlinear Methods

Summary

The various notions of absolute stability introduced in the previous chapter provided for a reasonable approach to the stiff problem. However, the restrictions imposed by these notions on the class of linear multistep methods (*e.g.*, loss of explicitness or limitations on accuracy) are severe indeed. Each weakening of A-stability, the strongest of these notions of absolute stability, gains some ground for the linear multistep class. Here we consider an alternative approach, namely, maintain A-stability but leave the class of linear multistep methods. We discuss two families of the many such possible approaches. First we consider a family of interpolatory methods which were in fact devised expressly for use on the stiff problem. This family is composed of methods by Certaine and by Jain (see Certaine, 1960, and Jain, 1972). Then we consider the venerable class of Runge–Kutta methods and point out that certain of these methods, including a variation of them by Rosenbrock (see Rosenbrock, 1962),are useful for the stiff problem.

3.1. INTERPOLATORY METHODS

We begin our discussion of interpolatory methods by describing Certaine's method and conclude it, following that, with a description of Jain's method.

3.1.1. *Certaine's Method*

The system of differential equations is written in the form

$$y'(t) = - Dy(t) + g(y(t), t). \tag{3.1.1}$$

Here y and g are m-vectors and D is an $m \times m$ constant matrix with at

43

least one large eigenvalue. We integrate (3.1.1) to obtain

$$y(t_{n+1}) = e^{-Dh}y_n + \int_{t_n}^{t_{n+1}} e^{D(t-t_{n+1})}g(y,t)dt, \qquad h = t_{n+1} - t_n.$$

$$(3.1.2)$$

Certaine's method consists of the following two steps.

(i) Approximate $g(y, t)$ by an interpolation polynomial, $g_k(t)$, of degree k at the points $t_{n-k}, t_{n-k+1}, \ldots, t_n$. Replace g in (3.1.2) by g_k, and use the resulting expression for $y(t_{n+1})$ as a predictor.

(ii) Using the predicted value of $y(t_{n+1})$, repeat step (i) using the points $t_{n-k+1}, \ldots, t_{n+1}$ to determine the correction.

Thus, Certaine's method is given by two utilizations of the following expression

$$y_{n+1} = e^{-Dh}y_n + e^{Dt_{n+1}} \int_{t_n}^{t_{n+1}} e^{Dt}g_k(t)dt, \qquad (3.1.3)$$

and it is apparent that we have left the linear multistep class.

Since some details concerning Certaine's method are extractable from the discussion of Jain's method to follow, we conclude our discussion of Certaine's method with the following two observations.

REMARK 3.1.1. The integral in (3.1.3) may be evaluated explicitly. If the exponential matrix e^{-D} is difficult to evaluate, one may take $D = D_1 + D_2$, where e^{-D_1} is easy to evaluate and $e^{D_2 t}$ is adjoined to g.

REMARK 3.1.2. If g is a polynomial or order less than $k + 1$, Certaine's method is exact. Thus the method is A-stable.

Remark 3.1.2 can be demonstrated along the lines of the proof of Theorem 3.1.3 in the next section.

3.1.2. *Jain's Method*

We start with the initial value problem

$$y'(t) = f(t, y), \qquad t \in (a, b).$$
$$y(a) = s.$$
$$(3.1.4)$$

Here y, s and f are m-vectors.

We consider the function

$$y'(t) + Py(t),$$

where P is an $m \times m$ matrix to be specified, and we perform the following three steps.

(i) Approximate $y'(t) + Py(t)$ by a polynomial of interpolation, $Q(t)$, which uses *Hermite interpolatory data* at the points $t_{n-i}, i = 0, 1, \ldots, k - 1$.

(ii) Integrate the differential equation $y' + Py = Q$ from t_n to t_{n+1}.

(iii) Choose P as an approximation to

$$-\left(\frac{\partial f}{\partial y}\right)_n \equiv -\frac{\partial f(t_n, y(t_n))}{\partial y}.$$

Step (i) results in

$$y'(t) + Py(t) = \sum_{i=1}^{k} h_i(t)(f_i + Py_i) + \sum_{i=1}^{k} \bar{h}_i(t)(f_i' + Pf_i) + T. \tag{3.1.5}$$

Here h_i and $\bar{h}_i$ are the fundamental Hermite interpolation polynomials of the first and second kind, respectively, corresponding to the points t_{n-i}, $i = 0, 1, \ldots, k - 1$. For clarity we do not display the dependence of the h_i and of the $\bar{h}_i$ on n. Also

$$f_i = f(t_i, y_i), \quad f_i' = f'(t_i, y_i),$$

$$T = \frac{1}{(2k)!} F^{(2k)}(\xi)\pi^2(t), \quad a < \xi < b,$$

where

$$F(t) = f(t) + Py(t)$$

and

$$\pi(t) = \prod_{i=0}^{k-1} (t - t_{n-i}).$$

Now we apply step (ii) (i.e., integrate (3.1.5)). We find

$$y_{n+1} = e^{-Ph}y_n + e^{-Pt_{n+1}} \sum_{i=1}^{k} [H_i F_i + \bar{H}_i F_i'] + R_n, \tag{3.1.6}$$

where

$$H_i = \int_{t_n}^{t_{n+1}} e^{Pt} h_i(t)dt,$$

$$\bar{H}_i = \int_{t_n}^{t_{n+1}} e^{Pt} \bar{h}_i(t)dt$$

and

$$R_n = \frac{e^{-Pt_{n+1}}}{(2k)!} \int_{t_n}^{t_{n+1}} e^{Pt} F^{(2k)}(\xi)\pi^2(\xi)dt.$$

As far as step (iii) is concerned and in the case where $m = 1$, a natural choice for P is

$$P = \frac{f_n - f(t_n, y_{n-1})}{y_n - y_{n-1}}.$$

In the case $m > 1$, the choices for P depend upon the relative dfficulty in evaluating e^{Ph}. A simple choice is the diagonal matrix whose iith entry is

$$P_{ii} = -\frac{f_n^i - f^i(t_n, y_n^1, \ldots, y_n^{i-1}, y_{n-1}^i, y_n^{i+1}, \ldots, y_n^m)}{y_n^i - y_{n-1}^i}.$$

As we see in (3.1.6), Jain's method is far from being a linear multistep method. Properties of this method are described in the following theorem.

THEOREM 3.1.3. *The method of Jain is A-stable and of order 2k.*
Proof. Let $f(t, y) = \lambda y$ where λ is a complex constant with Re $\lambda < 0$ (*i.e.*, the case of the test equation). Then $P = -\lambda$, and for each i,

$$F_i = f_i + Py_i = \lambda y_i - \lambda y_i = 0$$

and

$$F_i' = f_i' + Pf_i' = \lambda y_i' - \lambda y_i' = 0.$$

Then (3.1.6) becomes

$$y_{n+1} = e^{\lambda h} y_n.$$

Then since Re $\lambda < 0$, $\lim_{n \to \infty} y_n = 0$ for each fixed $h > 0$. This demonstrates the A-stability of the method.

Now insert $s = (t - t_n)/h$ into (3.1.6). It becomes

$$y_{n+1} = e^{-Ph} y_n + h e^{-Ph} \sum_{i=1}^{k} [H_i F_i + \bar{H}_i F_i'] + R_n. \qquad (3.1.7)$$

Here

$$H_i = \int_0^1 e^{Phs} k_i(s)ds, \quad k_i(s) = h_i(hs + t_i),$$

$$\bar{H}_i = \int_0^1 e^{Phs} \bar{k}_i(s)\,ds, \quad \bar{k}_i(s) = \bar{h}_i(hs + t_i)$$

and

$$R_n = \frac{h^{2k+1}}{2k!} e^{-Ph} \int_0^1 e^{Phs} F^{(2k)}(\xi)\pi^2(s)\,ds$$

$$= \frac{h^{2k+1}}{2k!} e^{-Ph} \int_0^1 F^{(2k)}(\xi)\pi^2(s)\,ds + O(h^{2k+2})$$

$$= \frac{h^{2k+1}}{2k!} e^{-Ph} F^{(2k)}(\bar{\xi}) \int_0^1 \pi^2(s)\,ds + O(h^{2k+2}),$$

by the second mean value theorem. Then

$$R_n = h^{2k+1} e^{-Ph} F^{(2k)}(\bar{\xi})\Lambda_k + O(h^{2k+2}),$$

where

$$\Lambda_k = \frac{1}{2k!} \int_0^1 \pi^2(s)\,ds.$$

Thus the method is order $2k$ and the theorem is proved. $\qquad\square$

Some Special Cases
The integrals for the determination of the H_i, $\bar{H}_i$ and R_n are of the form

$$\int_0^1 e^{Phs}\left(\sum_{i=1}^N A_i s^i\right)ds,$$

where $N = N(n)$ is an integer. In addition

$$H_i = \sum_{r=1}^{2k} a_r(Ph)^{-r}e^{Ph} + h\sum_{r=1}^{2k} b_r(Ph)^{-r},$$

$$\bar{H}_i = \sum_{r=1}^{2k} \alpha_r(Ph)^{-r}e^{Ph} + h\sum_{r=1}^{2k} \beta_r(Ph)^{-r}$$

In the simple case $k = 1$, we find

$$h_1(t) = 1, \qquad \bar{h}_1(t) = t - t_1,$$

$$k_1(s) = 1, \qquad \bar{k}_1(s) = s, \qquad \pi(s) = s,$$

$$H_1 = \int_0^1 e^{Phs}\,ds = (Ph)^{-1}(e^{Ph} - 1),$$

$$\bar{H}_1 = \int_0^1 s e^{Phs} ds = [(Ph)^{-1} - (Ph)^{-2}] e^{Ph} - (Ph)^{-2},$$

$$\Lambda_1 = \tfrac{1}{2} \int_0^1 s^2 ds = \tfrac{1}{6},$$

$$a_1 = 1, \qquad a_2 = 0, \qquad b_1 = -1, \qquad b_2 = 0,$$
$$\alpha_1 = 1, \qquad \alpha_2 = -1, \qquad \beta_1 = 0, \qquad \beta_2 = 1.$$

We conclude this section with the following remark.

REMARK 3.1.4. While the methods of Certaine and of Jain are A-stable and of higher accuracy, they are computationally costly to use.

3.2. RUNGE–KUTTA METHODS AND ROSENBROCK METHODS

In this section, we discuss the well-known class of Runge–Kutta methods, and show that in this class of methods we may also find A-stable methods of higher order. Then we consider a variant of these methods due to Rosenbrock which have desirable computational properties.

3.2.1. Runge-Kutta Methods with v-levels

We start with the differential equation

$$\dot{x} = f(x), \tag{3.2.1}$$

where x and f are m-vectors.

A Runge–Kutta process with v levels is defined by the following relations.

(a) $\quad x^+ = x + h \sum_{i=1}^{v} b_i k_i$

$$\tag{3.2.2}$$

(b) $\quad k_i = f(x + h \sum_{j=1}^{v} a_{ij} k_j), \qquad i = 1, 2, \ldots, v.$

These relations are used to define an approximation, x^+ to $x(t_{n+1})$ in terms of an approximation to $x(t_n)$, denoted simply by x in (3.2.2). The coefficients $b_i, a_{ij}, i, j = 1, 2, \ldots, v$ are to be determined by a procedure which we now describe.

3.2.2. Determination of the Coefficients

By using (3.2.1), we may write the following list of formal relations

$$x^{(1)} = f,$$
$$x^{(2)} = f_1 f,$$
$$x^{(3)} = f_2 f^2 + f_1^2 f,$$
$$x^{(4)} = f_3 f^3 + 3(f_2 f)(f_1 f),$$

$$\qquad (3.2.3)$$

$$\cdots\cdots$$

$$x^{(r)} = \sum_{s=1}^{p} \alpha_{rs} F_{rs}.$$

Here $f_1 = f_x$, the Jacobian, an array or order 2, $f_2 = f_{xx}$, the Hessian, an array of order 3, ...

The F_{rs}, $r = 1, 2, \ldots, s = 1, \ldots, p$, are called the elementary differentials. For each index r, there are p_r such differentials. For example, $p_1 = 1$, $p_2 = 1, p_3 = 2, p_4 = 4, \ldots$, and

$$F_{11} = f, \quad F_{21} = f_1 f, \quad F_{31} = f_2 f^2, \quad F_{32} = f_{32} = f_1^2 f.$$

Now let x^+ and x denote the exact value of x at t_{n+1} and t_n respectively. Next substituting the relations in (3.2.3) into the formal statement $x^+ - x = \sum\limits_{r=1}^{\infty} h^r x^{(r)}/r!$, of Taylor's theorem gives

$$x^+ - x = \sum_{r=1}^{\infty} \frac{1}{r!} h^r \left(\sum_{s=1}^{p_r} \alpha_{rs} F_{rs} \right). \qquad (3.2.4)$$

Now if we formally develop each k_i, $i = 1, \ldots, v$ (see (3.2.2b)) in a series, we may write the first relation in (3.2.2) as

$$h \sum_{i=1}^{v} b_i k_i = \sum_{r=1}^{\infty} \frac{1}{(r-1)!} h^r \left(\sum_{s=1}^{p_r} \beta_{rs} \phi_{rs} F_{rs} \right). \qquad (3.2.5)$$

Here the β_{rs} are numerical coefficients while the ϕ_{rs} are functions of the b_i and the a_{ij}.

For a Runge–Kutta process to be of order or precision p, it is necessary that the formal series in (3.2.4) and (3.2.5) agree to p terms. Thus, we find

$$\phi_{rs} = \alpha_{rs}/(r\beta_{rs}), \qquad r = 1, \ldots, p, \qquad s = 1, \ldots, p_r, \qquad (3.2.6)$$

as a set of $M = \sum_{r=1}^{p} p_r$ equations for the determination of the $v(v + 1)$ coefficients $a_i, b_{ij}, i, j = 1, \ldots, v$.

We distinguish three classes of Runge–Kutta processes as prescribed in the following definition.

DEFINITION 3.2.1. A Runge–Kutta process is said to be explicit if $a_{ij} = 0, j \geq i$, it is said to be semi-explicit if $a_{ij} = 0, j > i$ and it is said to be implicit otherwise. The number of available coefficients in these three cases are N_e, N_s and N_i, respectively, where

$$N_e = v(v + 1)/2, \qquad N_s = v(v + 3)/2, \qquad N_i = v(v + 1).$$

The relation between the quantities N_e, N_s, N_i, p and M for $v = 1, \ldots, 7$ is expressed in the following Table 3.2-1.

TABLE 3.2-1

v	N_e	N_s	N_i	p	M
1	1	2	2	1	1
2	3	5	6	2	2
3	6	9	12	3	4
4	10	14	20	4	8
5	15	20	30	5	17
6	21	27	42	6	37
7	28	35	56	7	85

The M equations in (3.2.6) are not independent, and so it is usually possible to satisfy them with a number N of coefficients considerably smaller than M.

3.2.3. An Example

Let us illustrate the last point by means of the case $p = v = 3$. In this case, an explicit calculation using (3.2.1) gives

$$h \sum_{i=1}^{3} b_i k_i = h \left(\sum_{i=1}^{3} b_i \right) F_{11} + h^2 \left(\sum_{i=1}^{3} b_i c_i \right) F_{21}$$

$$+ \frac{h^3}{2} \left[\left(\sum_{i=1}^{3} b_i c_i^2 \right) F_{31} + 2 \left(\sum_{i=1}^{3} \sum_{j=1}^{3} b_i a_{ij} c_j \right) F_{32} \right] + O(h^4), \qquad (3.2.7)$$

where

$$c_i = \sum_{j=1}^{3} a_{ij}.$$

This must be set equal to the right member of (3.2.5) which is

$$h(\beta_{11}\phi_{11})F_{ii} + h^2(\beta_{21}\phi_{21})F_{21}$$

$$+ \frac{h^3}{2}(\beta_{31}\phi_{31}F_{31} + \beta_{32}\phi_{32}F_{32}) + O(h^4). \tag{3.2.8}$$

Comparing coefficients of the elementary differentials in (3.2.7) and (3.2.8) allows us to determine $\beta_{rs}\phi_{rs}$ as functions of the a_{ij} and the b_i. These are:

$$\beta_{11}\phi_{11} = \sum_{i=1}^{3} b_i,$$

$$\beta_{21}\phi_{21} = \sum_{i=1}^{3} b_i c_i,$$

$$\beta_{31}\phi_{31} = \sum_{i=1}^{3} b_i c_i^2, \tag{3.2.9}$$

$$\beta_{32}\phi_{32} = 2\sum_{i=1}^{3}\sum_{j=1}^{3} b_i a_{ij} c_j.$$

Next, the expression in (3.2.4) must be developed so that the α_{rs} may be obtained. This reveals that $\alpha_{11} = 1$, $\alpha_{21} = 1$, $\alpha_{31} = 1$ and $\alpha_{32} = 1$. (Recall that we have already noted that $p_1 = p_2 = 1$ and $p_3 = 2$.)

We assemble the information developed for this example in Table 3.2-2.

TABLE 3.2-2

$\beta\phi = \alpha/r$	r	α	p_r
$\beta_{11}\sum b_i = 1$	1	1	1
$\beta_{21}\sum b_i c_i = 1/2$	2	1	1
$\beta_{31}\sum b_i c_i^2 = 1/3$	3	1 } 2	
$\beta_{32}\sum_i\sum_j b_i a_{ij} c_i = 1/3$	3	1	

TABLE 3.2-3

$a_{11} \ldots \ldots a_{1v}$	c_1	
$\ldots \ldots$		
$a_{v1} \ldots \ldots a_{vv}$	c_v	where $\quad c_i = \sum\limits_{j=1}^{v} a_{ij}$
$b_1 \ldots \ldots b_v$		

One associates the tableau of coefficients in Table 3.2-3 with the Runge–Kutta process.

A particular solution of the equations displayed in Table 3.2-2 is displayed in the version of Table 3.2-3 corresponding to $v = 3$ as follows.

TABLE 3.2-4

0	0	0	0
1/2	0	0	1/2
-1	2	0	1
1/6	2/3	1/6	

The corresponding values of the β's are $\beta_{11} = \beta_{21} = \beta_{31} = 1$ and $\beta_{32} = 2$. This particular solution is due to Kutta.

For a more detailed treatment of the derivation of Runge–Kutta methods, see Butcher, 1964.

3.2.4. Semi-explicit Processes and the Method of Rosenbrock

Among the implicit and semi-explicit Runge–Kutta processes (see Definition 3.2.1) are A-stable methods. The implicit processes lead to methods which are difficult to apply in general, because at each step of the integration, the k_i, $i = 1, \ldots, v$ must be determined as the solution of the system of v nonlinear equations (3.2.2b).

In the semi-explicit case, the nonlinear system is triangular in the sense that the jth equation in this system contains only the unknowns k_i, $i = 1, \ldots, j$. Thus, each equation in turn need only be solved for one unknown, i.e., the ith equation for k_i, $i = 1, \ldots, v$.

Let us consider the semi-explicit case and replace the solution procedure for the k_i, $i = 1, \ldots, v$ by a single step of a Newton–Raphson iteration.

The resulting method is

$$x^+ = x + h \sum_{i=1}^{\nu} b_i k_i, \tag{3.2.10}$$

$$k_i = \left[I - h a_{ii} f_x \left(x + h \sum_{j=1}^{i-1} a_{ij} k_j \right) \right]^{-1} f \left(x + h \sum_{j=1}^{i-1} c_{ij} k_j \right), \quad i = 1, \dots, \nu,$$

where I is the $m \times m$ identity matrix. This is an example of a method which may be called a linearized semi-explicit Runge–Kutta process of the Rosenbrock type, or simply a Rosenbrock method. (See Rosenbrock, 1962.)

Using Rosenbrock's notation, the case $p = 3$, $\nu = 2$ becomes

$$\begin{aligned}
x^+ &= x + h(R_1 k_1 + R_1 k_2), \\
k_1 &= [I - h a_1 f_1]^{-1} f, \\
k_2 &= [I - h a_2 f_1 (x + h c_1 k_1)]^{-1} f(x + h b_1 k_1).
\end{aligned} \tag{3.2.11}$$

There are six undetermined coefficients. The set of equations analogous to (3.2.6) for the determination of the six unknowns are four in number and are

$$\begin{aligned}
R_1 + R_2 &= 1, \\
R_1 a_1 + R_2 (a_2 + b_1) &= \tfrac{1}{2}, \\
R_1 a_1^2 + R_2 [a_2^2 + (a_1 + a_2) b_1] &= \tfrac{1}{6}, \\
R_2 (a_2 c_1 + \tfrac{1}{2} b_1^2) &= \tfrac{1}{6}.
\end{aligned} \tag{3.2.12}$$

A particular solution of (3.2.12) due to Rosenbrock is

$$\begin{aligned}
a_1 &= 1 + 1/\sqrt{6}, \\
a_2 &= 1 - 1/\sqrt{6}, \\
b_1 &= c_1 = [-6 - \sqrt{6} + (58 + 20\sqrt{6})^{1/2}]/(6 + 2\sqrt{6}), \\
R_1 &= -0.413154, \\
R_2 &= -1.413154.
\end{aligned}$$

The two matrices in (3.2.11) which must be inverted become identical under the constraints $a_1 = a_2$ and $c_1 = 0$. This considerably reduces the computation per step. Under these constraints the equations (3.2.12) become

$$R_1 + R_2 = 1,$$
$$a_1 + R_2 b_1 = \tfrac{1}{2},$$
$$a_1^2 + 2R_2 a_1 b_1 = \tfrac{1}{6},$$
$$R_2 b_1^2 = \tfrac{1}{3}.$$

(3.2.13)

(3.2.13) has two solutions, one of which is

$$R_1 = 3/4, \quad R_2 = 1/4, \quad a_1 = (1 + \sqrt{3})/2, \quad b_1 = -2/\sqrt{3}.$$

(See Calahan, 1967 for a study of this solution of (3.2.13).)

3.2.5. A-Stability

To demonstrate the A-stability of these linearized methods requires their application to the scalar test equation (viz. $f = \lambda x, f_x = \lambda$) and a study of the location of the roots of the characteristic equation corresponding to the difference equation which results. We forego these details.

Chapter 4

Exponential Fitting

Summary

We have now completed a review of some of the ideas and methods for approximating the solution of stiff equations which use a technique coupling small mesh increments during a transitory stage with a property of absolute stability during a permanent stage. We now turn to a second class of methods which employ exponential fitting, a different approach to the stiff problem.

In the context of a simple example, we see in Section 1.1.1 that the control of the error $e_n = u_n - y_n$ (see 1.1.6) depends on the stability of the amplification operator, $K(hA)$, and the closeness of $K(hA)$ to the solution operator, $S(hA)$. We see in that example that $K(hA)$ is made close to $S(hA)$ by making $K(hz)$ close to $S(hz)$ for z in the spectrum $\sigma(A)$ of A. This in turn is accomplished by making K close to S in a neighborhood of the origin, and then shrinking $h\sigma(A)$ into this neighborhood by making h small enough.

The methods of exponential fitting replace the single point at the origin by a set of points in the complex plane, which we call the fitting points. Then $K(z)$ is made close to $S(z)$ at all points in this set. Then by taking h small, the collection of points $h\sigma(A)$ tend to one or another of the fitting points.

This idea becomes interesting for stiff systems when we note that fitting points may be very large in magnitude, so that h is not required to scale the entire spectrum of A into a neighborhood of the origin. Of course, in addition to being fitted, a method must be stable and convergent in some sense if it is to be of computational value. We discuss these latter aspects as well. Exponential fitting is effective in the transient stage where the rapidly varying modes of the solution are present. When these modes become quiescent in the permanent stage, the value of fitting is more or less lost and thus it becomes unnecessary. In the permanent stage, the rapidly varying modes must be kept quiescent by use of a method with some form of absolute stability. Of course in practice, the quiescent

55

modes will typically tend to reactivate, and the fitting may need to be reperformed from time to time according to some adaptive criteria.

We begin in Section 4.1 with some examples of exponential fitting for linear multistep methods and include as well an example of an error analysis of a particular exponentially fitted method, the Willoughby-Liniger–Miranker method. In Section 4.2, we generalize these ideas to consider fitting for a class of matricial linear multistep methods. In Section 4.3, we consider the highly oscillatory problem wherein the rapidly varying modes correspond to eigenvalues with large imaginary parts. For these problems the transitory stage is itself permanent, and we consider fitting in such a case. Finally, in Section 4.4, we consider some numerical approaches to partial differential equations which employ exponential fitting.

4.1. EXPONENTIAL FITTING FOR LINEAR MULTISTEP METHODS

4.1.1. *Motivation and Examples*

We motivate the idea of *exponential fitting* by means of several examples. Consider the following linear multistep formulas (4.1.1)–(4.1.4).

$$F_1: x_{n+1} - x_n - h[(1-a)\dot{x}_{n+1} + a\dot{x}_n] = 0. \tag{4.1.1}$$

The order of this method is $p = 2$ if $a = 1/2$ and it is $p = 1$ otherwise.

$$F_2: x_{n+1} - x_n - \frac{1}{2}h[(1+a)\dot{x}_{n+1} + (1-a)\dot{x}_n)]$$

$$+ \frac{1}{4}h^2[(b+a)\ddot{x}_{n+1} - (b-a)\ddot{x}_n] = 0. \tag{4.1.2}$$

Here $p = 4$ if $b = \frac{1}{3}$ and $a = 0$, $p = 3$ if $b = \frac{1}{3}$, $a \neq 0$ and $p = 2$ if $b \neq \frac{1}{3}$. In particular for $b = \frac{1}{3}$, (4.1.2) becomes

$$F_3: x_{n+1} - x_n - \frac{h}{2}[(1+a)\dot{x}_{n+1} + (1-a)\dot{x}_n]$$

$$+ \frac{h^2}{12}[(1+3a)\ddot{x}_{n+1} - (1-3a)\ddot{x}_n] = 0. \tag{4.1.3}$$

In turn, when $a = 0$, (4.1.3) becomes

$$F_4: x_{n+1} - x_n - \frac{h}{12}(\dot{x}_{n+1} + \dot{x}_n) + \frac{h^2}{12}(\ddot{x}_{n+1} - \ddot{x}_n) = 0. \tag{4.1.4}$$

(4.1.2)–(4.1.4) are not the customary linear multistep methods since they employ second derivatives of x.

The exact solution of the test equation (see (2.1.2)) satisfies the following recurrence relation

$$x(t_{n+1}) = e^q x(t_n), \qquad q = \lambda h. \tag{4.1.5}$$

The amplification factor of F_v is $K_v(q)$, $v = 1, 2, 3, 4$ (see (1.1.5) f.) where

$$
\begin{aligned}
K_1(q) &= (1 + aq)/[1 - (1 - a)q], \\
K_2(q) &= [4 + 2(1 - a)q + (b - a)q^2]/[4 - 2(1 + a)q + (b + a)q^2], \\
K_3(q) &= [12 + 6(1 - a)q + (1 - 3a)q^2]/[12 - 6(1 - a)q \\
&\quad + (1 + 3a)q^2], \\
K_4(q) &= [12 + 6q + q^2]/[12 + 6q + q^2].
\end{aligned}
\tag{4.1.6}
$$

It is simple matter to verify that the truncation operator (see (1.1.6)f.)

$$T_v(q) = K_v(q) - e^q = O(q^{v+1}) \tag{4.1.7}$$

as $q \to 0$, since, as we have noted, p has the various values 2, 3, or 4 as the case may be.

We introduce the following definition of exponential fitting.

DEFINITION 4.1.1. A method with truncation operator $T(q)$ is *exponently fitted* to order r at a point c if $(d^j/dq^j) T(q)|_{q=c} = 0, j = 0, 1, \ldots, r$.

We note that the formulas F_v are exponentially fitted to order $r \geq v$ at the origin. The remaining parameters may be chosen so that fitting occurs elsewhere as well. If we can adjust F_v so that $T_v(h\lambda) = 0$, where the magnitude of λ is very large, then it is reasonable to use F_v to solve stiff systems whose spectrum is divided into two clusters. The first cluster lying near $q = 0$ corresponds to slowly varying modes; the second cluster, lying near $q = h\lambda = c$, corresponds to rapidly varying (stiff) modes.

Let us now consider some fittings of the F_v.

For $a = 0$, F_1 is fitted to order $r = 0$ at $c = -\infty$.

For $a = \frac{1}{2}$, F_1 becomes the trapezoidal formula. The fitting is maximal at $q = 0(p = r = 2)$, but there is no fitting at $c = -\infty$, since $\lim_{q \to -\infty} T_1(q) = -1$.

For $v = 1$ or 3, $T_v(c) = 0$ defines the parameter a as a function $a = a_v(c)$, where

$$a_1(q) = -q^{-1} - (e^{-q} - 1)^{-1},$$
$$a_3(q) = \tfrac{1}{3}[12 + 6q + q^2 - (12 - 6 + q^2)e^q]/[2q + q^2$$
$$- (2q - q^2)e^q].$$

(4.1.8)

$T_2(c) = T_2(c') = 0$ defines a and b as functions of both c and c'. These two functions are

$$a_2(q, q') = 2[f(q) - f(q')]/[q'f(q) - qf(q')]$$

and

$$b_2(q, q') = 2(q' - q)[q'f(q) - qf(q')],$$

respectively. Here

$$f(q) = q^2(e^q - 1)/[2 + q + (q - 2)e^q].$$

4.1.2. Minimax Fitting

As an alternate use of free parameters, we may attempt to minimize $T(q)$ in some global sense. We illustrate this by means of the following example dealing with F_1.
Let

$$\bar{T}(a) = \max_{-\infty < q \leq 0} |T(q)|.$$

The following lemma results from a direct calculation which employs (4.1.8). (See Liniger and Willoughby, 1970.)

LEMMA 4.1.2. $a = a_1(c)$ defines a 1-1 mapping of $c \in (-\infty, 0]$ into $a \in [0, \tfrac{1}{2}]$.

Now let a_0 be defined by

$$\bar{T}(a_0) = \min_{0 \leq a \leq 1/2} \bar{T}(a) = \min_{-\infty < c \leq 0} \bar{T}(a_1(c)).$$

Then

$$a_0 = 0.122\ldots, \qquad \bar{T}(a_0) = 0.139\ldots,$$

and the corresponding fitting point is $c_0 = -8.19\ldots$. Notice that for the backward Euler formula $\bar{T}(0) = 0.204\ldots$, while $\bar{T}(1/2) = 1$ for the trapezoidal formula.

4.1.3. *An Error Analysis for an Exponentially Fitted* F_1

In the classical case, fitting at the origin is a form of control of the local error, i.e., is tantamount to what we call local error analysis. Then we see that exponential fitting is a somewhat elaborated variant of local error analysis. Just as in the classical procedure wherein a local error analysis by no means assures the control of the global error, we also lack this assurance in the case of exponential fitting. We must supplement the local analysis with a stability analysis, and then combine the two by constructing a global error analysis to demonstrate the value of the method.

We now illustrate such a global error analysis with F_1 (see (4.1.1)). In Section 4.2, we will consider a more general framework.

When F_1 is applied to the linear system (1.1.1), *viz*

$$\dot{y} = Ay, \tag{4.1.9}$$

we find the following recurrence relation for the global error, e_n (see (4.1.6) and (4.1.7)).

$$e_{n+1} = K_1(hA)e_n + T_1(hA)y_n. \tag{4.1.10}$$

From this in turn, we get

$$e_n = \sum_{j=0}^{n-1} K_1^j(hA)T_1(hA)y_{n-j-1}, \tag{4.1.11}$$

where we have assumed that the initial error, $e_0 = 0$.

The following lemma follows from a direct calculation.

LEMMA 4.1.3. $|K_1(z)| < 1$ *for* $z \in (0, -\infty)$ *and* $a \in [0, \frac{1}{2}]$.

This lemma implies that F_1 is A-stable for $a \in [0, 1/2]$ (see Lemma 4.1.2). We now consider a to be restricted to this interval.

Now let us suppose that A is negative definite and has distinct eigenvalues λ_i, $i = 1, \ldots, m$, where $0 > \lambda_1 > \ldots > \lambda_m$. Let the resolution of the identity, relative to A be given by

$$I = \sum_{i=1}^{m} P_i(A), \tag{4.1.12}$$

where the P_i, $i = 1, \ldots, m$ are appropriate polynomials (see (1.1.10)f.).

Then

$$\| K_1^j(hA)\| = \| \sum_{i=1}^{m} K_1^j(h\lambda_i)P_i(A)\|$$

$$\leq \text{const} \times \sum_{i=1}^{m} |K_1(h\lambda_i)| \leq \text{const}. \tag{4.1.13}$$

The first equality in (4.1.13) follows from (4.1.12) while the last inequality follows from Lemma 4.1.3, since the λ_i are negative. Using (4.1.13), (4.1.11) becomes

$$\|e_n\| \leq \text{const} \times n \| T_1(hA)\|. \tag{4.1.14}$$

Now from the properties of $T_1(z)$ for z near zero, we may conclude that

$$|T_1(z)| \leq \text{const} \times \min(1, z^2), \qquad z \leq 0. \tag{4.1.15}$$

On the other hand, given $c > 0$, and if $a = a_1(c)$ (see (4.1.8)), then from Taylor's theorem, we conclude that

$$T_1(z) = (c - z)(K_1'(\tilde{z}) + e^{\tilde{z}}). \tag{4.1.16}$$

Here $\tilde{z} \in (c, z)$ arises from the remainder term in the application of Taylor's theorem. From (4.1.16) in turn, we have that

$$|T_1(z)| \leq \text{const} \times |c - z|, \qquad c < 0, \qquad z \leq 0. \tag{4.1.17}$$

Now let (I_1, I_2) be a partition, Π, of the integers $J = \{1, \ldots, m\}$. Then combining (4.1.14)–(4.1.17) and utilizing the resolution of the identity, we get the following estimate for $\|e_n\|$.

$$\|e_n\| \leq n\,\text{const} \times \min_{\Pi} \left[\max_{i \in I_1} |h^2 \lambda_i^2| + \max_{i \in I_2} h|\gamma - \lambda_i| \right]$$

$$\leq \text{const} \times \max_{i \in J} \left[\min(|h\lambda_i|^2, |\gamma - \lambda_i|) \right].$$

(Recall that $c = h\gamma$.)

The property of Lemma 4.1.2 (i.e., the fitting) was observed by R. A. Willoughby while that of Lemma 4.1.3 (i.e., the A-stability) was observed by W. Liniger. The global error analysis was made by W. L. Miranker. Thus, the simple scheme F_1 used in an exponential fitting mode for approximating the solution of stiff equations is called the Willoughby–Liniger–Miranker method. (See Liniger and Willoughby, 1970 and Miranker, 1971b.)

4.2. FITTING IN THE MATRICIAL CASE

In this section, we study the process of exponential fitting in a setting which is more general than that of Section 4.1. In particular, we consider a class of linear multistep methods with matricial coefficients. Such a general setting creates more possibilities for the use of fitting to deal with the stiff problem.

4.2.1. *The Matricial Multistep Method*

We consider the initial value problem for the following system.

$$\dot{x} = Ax, \qquad t > 0. \tag{4.2.1}$$

Here x is an m-vector and A is an $m \times m$ matrix of constants. Evidently

$$x_n = e^{Ah} x_{n-1}. \tag{4.2.2}$$

Now consider the three functions $L(z)$, $R(z)$ and $C(z)$ given as follows.

$$L(z) = \sum_{j=0}^{r} (\alpha_j + z\beta_j)e^{(r-j)z},$$

$$R(z) = \sum_{j=0}^{r} (\gamma_j + z\delta_j)e^{(r-j)z}, \tag{4.2.3}$$

$$C(z) = L(z)[R(z)]^{-1}.$$

Here the $\alpha_j, \beta_j, \gamma_j$ and $\delta_j, j = 0, \ldots, r$ are $m \times m$ matrices. Note that

$$L(hA) - C(hA)R(hA) \equiv 0. \tag{4.2.4}$$

Let $P(z)$ be an approximation to $C(z)$, and consider the following formula, which is an approximation to (4.2.4), as a numerical method for determining u_n as an approximation to $x_n, n = r, r+1, \ldots$.

$$\sum_{j=0}^{r} \alpha_j u_{n-j} + h \sum_{j=0}^{r} \beta_j A u_{n-j}$$

$$- P(hA)\left[\sum_{j=0}^{r} \gamma_j u_{n-j} + h \sum_{j=0}^{r} \delta_j A u_{n-j} \right] = 0. \tag{4.2.5}$$

If $P(z)$ were equal to $C(z)$, this expression would be an identity for solutions of (4.2.1) (see (4.2.4)). That is, (4.2.5) would be fitted (exponentially) at all points in the spectrum $\sigma(A)$. However, $C(hA)$ is too difficult to calculate, especially if we use (4.2.5) on systems of the form (4.2.1), where A changes at each step. Thus, we will choose $P(z)$ as a function for which $P(hA)$ is easy to calculate, and such that $P(z)$ is an approximation to $C(z)$ in a sense to be made precise.

4.2.2. The Error Equation

To determine the quality of (4.2.5) as a numerical method, we proceed to derive an equation for the global error $e_n = u_n - x_n$. To do this, we introduce the shift operator H, where

$$Hf(t) = f(t + h), \tag{4.2.6}$$

and we introduce two operators $\mathscr{L}(H)$ and $\mathscr{R}(H)$ associated respectively with L and R as follows.

$$\mathscr{L}(H) = \sum_{j=0}^{r} (\alpha_j + hA\beta_j)H^{r-j},$$

$$\mathscr{R}(H) = \sum_{j=0}^{r} (\gamma_j + hA\delta_j)H^{r-j}. \tag{4.2.7}$$

(Expect for the sign change, $\beta_j \to -\beta_j$, the $\mathscr{L}$ here is the same as the one used in Section 1.2.)

Now

$$Hx = e^{hA}x, \tag{4.2.8}$$

where x is a solution of (4.2.1).
Thus

$$(HA - AH)x = 0. \tag{4.2.9}$$

From this we deduce that

$$\mathscr{R}(H)x = R(hA)x,$$
$$\mathscr{L}(H)x = L(hA)x \tag{4.2.10}$$

and that

$$[\mathscr{L}(H) - C(hA)\mathscr{R}(H)]x_{n-r} = 0, \qquad n = r, r+1, \dots. \tag{4.2.11}$$

On the other hand, we may write (4.2.5) as

$$|\mathscr{L}(H) - P(hA)\mathscr{R}(H)]u_{n-r} = 0, \qquad n = r, r+1, \dots. \tag{4.2.12}$$

Then by subtracting (4.2.11) from (4.2.12), we find the following error equation, for $e_n = u_n - x_n, n = 0, 1, \ldots$.

$$[\mathscr{L}(H) - P(hA)\mathscr{R}(H)]e_{n-r}$$
$$= [P(hA) - C(hA)]\mathscr{R}(H)x_{n-r}, \qquad n = r, r+1, \ldots. \quad (4.2.13)$$

4.2.3. Solution of the Error Equation

To solve (4.2.13), we introduce the operator $\mathscr{S}(H)$ as follows.

$$\mathscr{S}(H) = \mathscr{L}(H) - P(hA)\mathscr{R}(H). \qquad (4.2.14)$$

We write $\mathscr{S}(H)$ as a polynomial in H as follows.

$$\mathscr{S}(H) = \sum_{j=0}^{r} s_j H^{r-j}, \qquad (4.2.15)$$

where

$$s_j \equiv s_j(A) \equiv \alpha_j + hA\beta_j - P(hA)(\gamma_j + hA\delta_j), \qquad j = 0, \ldots, r. \qquad (4.2.16)$$

Thus (4.2.13) may be written in the following form

$$\mathscr{S}(H)e_{n-r} = [P(hA) - C(hA)]\mathscr{R}(hA)x_{n-r}, \qquad n = r, r+1, \ldots. \qquad (4.2.17)$$

Now let

$$S(z) = \sum_{j=0}^{r} s_j z^{r-j} \qquad (4.2.18)$$

be a polynomial with the matricial coefficients $s_j, j = 0, \ldots, r$. Suppose that $[z^r S(z^{-1})]^{-1}$ is an analytic function of z in a neighborhood of $z = 0$ and let its power series be given by

$$[z^r S(z^{-1})]^{-1} = \sum_{j=0}^{\infty} \sigma_j z^j, \qquad (4.2.19)$$

where the σ_j are matrices. (See Lemma 4.2.2 below.)

Multiply (4.2.17) by σ_{N-n}, and sum the result over n from r to N. For the left member, this operation gives

$$\sum_{n=r}^{N} \sigma_{N-n} \mathscr{S}(H)e_{n-r} = \sum_{n=r}^{N} \sigma_{N-n} \sum_{j=0}^{r} s_j H^{r-j} e_{n-r}$$
$$= \sum_{n=r}^{N} \sigma_{N-n} \sum_{j=0}^{r} s_j e_{n-j} \qquad (4.2.20)$$

$$= \sigma_0 s_0 e_N + (\sigma_1 s_0 + \sigma_0 s_1)e_{N-1} + \ldots + (\sigma_{N-r} s_0 + \ldots + \sigma_n s_n)e_r + \text{linear combination of } e_0, e_1, \ldots, e_{r-1}.$$

From the defining property (4.2.19) of the $\sigma_j, j = 0, \ldots$, we may deduce the following relation.

$$\sum_{j=0}^{r} \sigma_{N-j} s_j = \delta_{N0} I. \tag{4.2.21}$$

Here I is the $m \times m$ identity matrix. Using (4.2.21) in (4.2.20) and assuming that the starting errors $e_0 = e_1 = \ldots = e_{r-1} = 0$, we find that the right member of (4.2.20) simply becomes e_N. Thus we are lead to the solution of (4.2.17), viz.

$$e_N = \sum_{n=r}^{N} \sigma_{N-n} [P(hA) - C(hA)] R(hA) x_{n-r}. \tag{4.2.22}$$

4.2.4. Estimate of Global Error

To estimate e_N, we require a stability statement and an accuracy statement. Stability is the subject of the following two lemmas.

LEMMA 4.2.1. *If $\sum\limits_{j=0}^{r} s_j(\lambda) z^{r-j}$ satisfies the root condition (see Definition 1.2.11) for each eigenvalue $\lambda \in \sigma(A)$, then the determinant $|S(z)|$ also satisfies the root condition.*

Proof. Let $f(A) = \sum\limits_{j=0}^{r} s_j(A) z^{r-j}$. Suppose that the determinant $|f(A)|$ vanishes for a value of z, then $|f(A) + \mu I - \mu I|$ vanishes. Then $\mu = \mu + f(\lambda)$, for each $\lambda \in \sigma(A)$ or $f(\lambda) = 0$ for that value of z. This completes the proof of the lemma. $\square$

LEMMA 4.2.2. *Let the determinant $|S(z)|$ obey the root condition. If the determinant of s_0 is not zero, then the matrix $[z^r S(z^{-1})]^{-1}$ is analytic in a neighborhood of $z = 0$. Furthermore, the matrices $\sigma_j, j = 0, 1, \ldots$ given by (4.2.19), have uniformly bounded norms.*

Proof. Since $z^r S(z^{-1}) = \sum\limits_{j=0}^{r} s_j z^j$ and $|s_0| \neq 0$, it is clear that $[z^r S(z^{-1})]^{-1}$ is analytic in a neighborhood of the origin. Since $|z^r S(z^{-1})| = z^{mr} |S(z^{-1})|$, the root condition locates the roots of the polynomial $|zS(z^{-1})|$ outside the open unit disc while those roots on the boundary of the unit disc are simple. Since

$$[z^r S(z^{-1})]^{-1} = [\text{matrix of polynomials in } z] / |z^r S(z^{-1})|,$$

it suffices to show that the power series for the reciprocal polynomial, $[z^r S(z^{-1})]^{-1}$ has bounded coefficients, given that its roots are outside the open unit disc, with those on the boundary being simple. Let $mr = q$, and let

$$|z^r S(z^{-1})|^{-1} = \left[\sum_{j=0}^{q} t_j z^j \right]^{-1} = \sum_{j=0}^{\infty} u_j z^j.$$

Then

$$u_n = \frac{1}{2\pi i} \oint \left(\zeta^{n+1} \sum_{j=0}^{q} t_j \zeta^j \right)^{-1} d\zeta,$$

where the contour of integration lies inside the unit disc and encircles the origin. If we move the contour through the unit disc and out to infinity in all directions, the integral will vanish if $q \geq 1$, and we are left with a sum of residues. If there is a pole ζ_0 on the unit disc, it is simple. Let the residue of this pole be τ_0. Then

$$|\tau_0| = \left| \left(\zeta_0^{n+1} \sum_{j=0}^{q} jt_j \zeta_0^{j-1} \right)^{-1} \right| = \left| \sum_{j=0}^{q} jt_j \zeta_0^{j-1} \right|^{-1},$$

which is independent of n.

If there is a pole at ζ_1 of order $\rho + 1$ ($\rho \geq 0$) outside the unit disc, let its residue be τ_1. Then

$$\tau_1 = \frac{d^\rho}{d\zeta^\rho} \left[(\zeta - \zeta_1)^{\rho+1} \left(\zeta^{n+1} \sum_{j=0}^{q} t_j \zeta^j \right)^{-1} \right]_{\zeta=\zeta_1}.$$

Let $Q(\zeta)$ be the polynomial given by

$$Q(\zeta) = \left(\sum_{j=0}^{q} t_j \zeta^j \right) \bigg/ (\zeta - \zeta_1)^{\rho+1}.$$

Then since $Q(\zeta)$ is independent of ζ_1,

$$\tau_1 = \frac{d^\rho}{d\zeta_1^\rho} [(\zeta_1^{n+1} Q(\zeta))^{-1}].$$

Then performing the differentiation, we get

$$\tau_1 = \sum_{j=0}^{\rho} \binom{\rho}{j} \left(\frac{d^j}{d\zeta_1^j} \zeta_1^{-n} \right) \frac{d^{\rho-j}}{d\zeta_1^{\rho-j}} (\zeta_1 Q(\zeta_1))^{-1}$$

$$= \sum_{j=0}^{\rho} (-1)^j \binom{\rho}{j} n(n+1)\ldots(n+j-1)\zeta_1^{-(n+j)} \frac{d^{\rho-j}}{d\zeta_1^{\rho-j}} (\zeta_1 Q(\zeta_1))^{-1}.$$

Thus

$$|\tau_1| \leq (n+\rho)^p K/|\zeta_1|^n,$$

where K is a constant independent of n. This estimate shows that $|\tau_1|$ tends to zero when n tends to infinity since $|\zeta_1| > 1$. Since there are at most a finite number of residues to be accounted for, the coefficients $u_n, n = 0, 1, \ldots,$ are bounded uniformly in n, and the lemma is proved. $\quad\square$

If $S(z)$ satisfies the hypothesis of Lemma 4.2.2, then that lemma and (4.2.22) may be combined to yield

$$\|e_N\| \leq \text{const} \times \|[P(hA) - C(hA)]R(hA)\| \sum_{n=r}^{N} \|x_{n-r}\|. \quad (4.2.23)$$

If $Nh = 1$, (4.2.23) becomes

$$\|e_N\| \leq \text{const} \times h^{-1} \|[P(hA) - C(hA)]R(hA)\|. \quad (4.2.24)$$

To complete the error analysis, the local error, which here is $\|[P(hA) - C(hA)]R(hA)\|$, must be made $o(h)$. To accomplish this we have at our disposal the specification of P, L and R to which we now turn.

4.2.5. Specification of P

Let $P(z)$ be a polynomial which has contact of order $\tau_i + 1$ with $C(z)$ at a set of points in the complex plane which we denote by $hz_i, i = 1, \ldots, p$. That is

$$P^{(j)}(hz_i) - C^{(j)}(hz_i) = 0, \qquad j = 0, 1, \ldots, \tau_i. \quad (4.2.25)$$

We suppose that $z_i \neq 0, i = 1, \ldots, p$, and we set $z_0 = 0$.

Now divide the spectrum of A into $p+1$ disjoint clusters called $k_0, \ldots, k_p$, respectively, where

$$k_i = \{\lambda_j \in \sigma(A) \mid |\lambda_j - z_i| \leq \min_{0 \leq l \leq p} |\lambda_j - z_l|\}.$$

Ties are decided arbitrarily.

Let

$$d_i = \max_{\lambda_j \in k_i} |\lambda_j - z_i|, \qquad i = 0, \ldots, p.$$

Now we resolve the identity by writing

$$I = \sum_{i=0}^{p} \sum_{\lambda_j \in k_i} Z_{ij}(hA), \quad (4.2.26)$$

where the Z_{ij} are appropriate polynomials (see $(1.1.10)f.$), and where for convenience, we have supposed that the eigenvalues of A are distinct. Using (4.2.26), we obtain

$$[P(hA) - C(hA)]R(hA) = \sum_{i=0}^{p} \sum_{\lambda_j \in k_i} [P(h\lambda_j)$$
$$- C(h\lambda_j)]R(h\lambda_j)Z_{ij}(hA). \qquad (4.2.27)$$

Using Taylor's theorem with remainder and (4.2.25), (4.2.27) becomes

$$[P(hA) - C(hA)]R(hA) = \sum_{\lambda_j \in k_0} [P(h\lambda_j)R(h\lambda_j) - L(h\lambda_j)]Z_{0j}(hA)$$
$$+ \sum_{i=1}^{p} \sum_{\lambda_j \in k_i} \frac{1}{\tau_i!}[h(\lambda_j - z_i)]^{\tau_i}[P^{(\tau_i)}(h\tilde{\lambda}_{ij})$$
$$- C^{(\tau_i)}(h\tilde{\tilde{\lambda}}_{ij})]R(h\lambda_j)Z_{ij}(hA). \qquad (4.2.28)$$

The $\tilde{\lambda}_{ij}$ and $\tilde{\tilde{\lambda}}_{ij}$ are values of λ arising in the remainder term.

4.2.6. Specification of L and R

To specify L and R we make the hypothesis

$$L(z) = O(z^{\mu+1}),$$
$$R(z) = O(z^{\nu+1}). \qquad (4.2.29)$$

This hypothesis asserts that the classical (matricial) linear multistep methods $\mathscr{L}(h)u_{n-1} = 0$ and $\mathscr{R}(H)u_{n-r} = 0$ have order of accuracy μ and ν, respectively, (see Definition 1.2.4).

Using (4.2.29) in (4.2.28) gives

$$\|[P(hA) - C(hA)]R(hA)\| \le c_1 \max(|hd_0|^{\nu+1}, |hd_0|^{\mu+1})$$
$$+ c_2 \sum_{i=1}^{p} \frac{1}{\tau_i!}|hd_i|^{\tau_i+1}. \qquad (4.2.30)$$

Here c_1 and c_2 are appropriate constants. (4.2.30) is the local error (estimate) for the numerical method (4.2.5) which we are studying. Combining (4.2.30) with (4.2.24) finally gives the global error estimate

$$\|e_N\| \le \text{const} \times \left[\max(|hd_0|^{\nu}, |hd_0|^{\mu}) + \sum_{i=1}^{p} |hd_i|^{\tau_i} \right]. \qquad (4.2.31)$$

The following observation connects the present development to the classical linear multistep theory.

REMARK 4.2.3. The classical theory of linear multistep methods corresponds to the case $P \equiv 0$.

4.2.7. *An Example*

A simple example of the method (4.2.5) corresponds to the case $r = 1$, $\alpha_0 = 1, \alpha_1 = -1$ and $\delta_1 = 1$. All other coefficients are taken to be zero. We select one cluster, i.e., $p = 1$, and $P(z)$ is taken to be the constant, $C(hz_1)$. The numerical method is

$$u_n - u_{n-1} = \frac{e^{hz_1} - 1}{hz_1} h\dot{u}_{n-1}. \tag{4.2.32}$$

For this method, $\mu = \nu = \tau_1 = 0$. Thus, the method has accuracy of order zero at the origin and at z_1. This low accuracy method may be viewed as the forward Euler method with mesh increment scaled by $(e^{hz_1} - 1)/(hz_1)$.

For this method, $S(z) = Iz - (I + ((e^{hz_1} - 1)/z_1)A)$. By Lemma 4.2.1, $|S(z)|$ obeys the root condition if $z - 1 - ((e^{hz_1} - 1)/z_1)\lambda$ does for every eigenvalue λ of A. This latter requirement is seen to be satisfied for any choice of z_1 in an interval which itself is contained in the interval $(-\infty, \lambda)$. (We suppose that $\lambda < 0$.) Thus if z_1 is chosen as any lower estimate for the spectrum of A, (4.2.32) will be stable.

Let us choose $z_1 = \min_{\lambda \in \sigma(A)} (\lambda - \mu)$, for some $\mu \geq 0$. To simplify for purposes of illustration, let us consider the special case corresponding to $m = 2$ and to say $\lambda_2 = -1$ and λ_1 some very large negative number. Then the difference scheme becomes

$$u_n - u_{n-1} = \frac{e^{h(\lambda_1 - \mu)} - 1}{\lambda_1 - \mu} A u_{n-1} \approx \frac{1}{\mu - \lambda_1} A u_{n-1},$$

since $\lambda_1 \ll -1$.

Now since $x_n = e^{Ah}x_{n-1}$ and $u_n = \left[I + \frac{e^{h(\lambda_1 - \mu)} - 1}{\lambda_1 - \mu} A \right] u_{n-1}$, we have

$$e_n = T(hA)e_{n-1},$$

where

$$T(hA) = I + \frac{e^{h(\lambda_1 - \mu)} - 1}{\lambda_1 - \mu} A - e^{Ah}.$$

Thus, $T(h\lambda)$ is the difference between the exponential $e^{\lambda h}$ and the

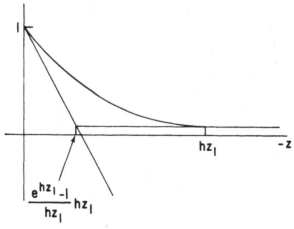

Fig. 4.2-1

straight line $1 - (e^{h(\lambda_1 - \mu)} - 1)\lambda/(\lambda_1 - \mu)$. At the eigenvalue λ_1, we have

$$T(h\lambda_1) = \frac{\mu}{\lambda_1} + e^{h\lambda_1}\left[1 + \mu\left(\frac{1}{\lambda_1} - h\right) + O\left(\mu^2\left(\frac{1}{\lambda_1} - h\right)^2\right)\right]$$
$$+ O\left(\frac{\mu^2}{\lambda_1^2}\right).$$

In Figure 4.2-1 we indicate how a forward Euler-type formula (viz. (4.2.32)) may be used to stably integrate a stiff system. From this figure, we see that the z-axis is scaled so that the method (the straight line) is used in a region where it is stable, but where its value (of the straight line) is equal to the value of the exponential (the transfer function of the solution) at the large eigenvalue.

We emphasize that the matricial class of methods being discussed here is very wide and the operative qualities of the class are by no means restricted to the scaling concept of the example.

4.3. EXPONENTIAL FITTING IN THE OSCILLATORY CASE

4.3.1. *Failure of Previous Methods*

The numerical methods which we discussed thus far use the fact that the rapid changes in the solution are transitory, although possibly recurrent on a time scale which is long compared to that of the rapid changes. When the stiff system has solutions of a *highly oscillatory character*, the methods which we have thus far examined do not work at all. For example, the key idea behind the introduction of notions of absolute stability was based on the existence of slowly varying stages in the development of the solution during which maintenance of the quiescence of the rapidly varying modes was the key idea.

In this section, we interpolate a discussion of a method for the highly oscillatory problem. This method employs a form of exponential fitting based on a process called aliasing (see Snider and Fleming, 1974).

4.3.2. Aliasing

Let $f(t)$ be periodic with period 2π. For a fixed integer $N > 0$, let the following values of $f(t)$ be given.

$$f(t_j), \qquad t_j = \left(\frac{j}{2N}\right) 2\pi, \qquad j = 0, 1, \ldots, 2N. \tag{4.3.1}$$

These points t_j are called the data points.

In terms of these values (4.3.1), the *discrete Fourier series* $C_N(t)$ of $f(t)$, is

$$C_N(t) = \frac{A_0}{2} + \sum_{r=1}^{N} (A_r \cos rt + B_r \sin rt) + \frac{A_N}{2} \cos Nt. \tag{4.3.2}$$

The coefficients of this series are

$$A_r = \frac{1}{N} \sum_{j=0}^{2N-1} f(t_j) \cos rt_j,$$

$$B_r = \frac{1}{N} \sum_{j=0}^{2N-1} f(t_j) \sin rt_j, \qquad r = 0, 1, \ldots, N.$$

If $f(t)$ is highly oscillatory, then a good representation of $f(t)$ by $C_N(t)$, requires N to be quite large. In fact we would need $2N$ values of $f(t)$ (see (4.3.1)) and $2N$ terms in the series (4.3.2), a large number of values and terms respectively.

Now suppose that $f(t)$ has a special form so that its frequencies fall into clusters. In particular, suppose that

$$f(t) = h(t) + \sum_{m=1}^{p} c_m \cos R_m t + d_m \sin R_m t,$$

where $h(t)$ is a smooth function. That is

$$h(t) = \frac{a_0'}{2} + \sum_{r=1}^{\infty} (a_r' \cos rt + b_r' \sin rt),$$

but that there exists an integer $L > 0$ such that the quantities $|a'_r|$ and $|b'_r|$ are negligible for $r > L$. Moreover, suppose that the frequencies $R_p > R_{p-1} > ... > R_1 > L$ are known (and are large).

The objective is to estimate the coefficients c_m and d_m, $m = 1, ..., p$ and the coefficients a'_r and $b'_r, r = 1, ..., L$. This may be efficiently accomplished through a process called *aliasing*.

Note that at each of the data points, the functions $\cos R_m t$ and $\sin R_m t$ can be replaced by $\cos r_m t$ and $\sin r_m t$, respectively, for some r_m where $R_m > N > r_m$. This is accomplished by use of the following identities.

$$\cos[2qN + r]t_j = \cos rt_j,$$
$$\cos[(2q + 1)N + r]t_j = -\cos(N - r)t_j,$$
$$\sin[2qN + r]t_j = \sin rt_j,$$
$$\sin[(2q + 1)N + r]t_j = -\sin(N - r)t_j.$$

One may view the first of these identities, for example, as the statement that $\cos[2qN + r]t$ takes on the same values as $\cos rt$ at the data points but oscillates faster in between. Thus, if we use a coarse mesh composed of $2N + 1$ mesh points where $N < R_1$, each of the high frequencies R_m will be replacable by a harmonic with an appropriate lower frequency $r_m < N$.

The relation between the Fourier coefficients (a_r, b_r) of $f(t)$ and the coefficients (A_r, B_r) of its finite Fourier series (see (4.3.2)) is

$$A_r = a_r + \sum_{m=1}^{\infty} (a_{2mN+r} - a_{2mN-r}),$$

$$(4.3.3)$$

$$B_r = b_r + \sum_{m=1}^{\infty} (b_{2mN+r} - b_{2mN-r}).$$

Thus the replacement of higher frequencies by lower ones will not mix and confuse components if N is chosen in such a way that each of the frequencies $\omega = 0, 1, 2, ..., L - 1, R_1, R_2, ..., R_p$ occurs in a separate sum in the right member of (4.3.3). Clearly, $N \geq L + p$, but usually N is smaller than R_p, making the process reasonably efficient.

4.3.3. *An Example of Aliasing*

These ideas are clearly illustrated with the following example. Suppose that $f(t)$ is the sum of a slowly varying function plus three harmonics of frequencies 177, 589 and 1000 respectively. Using $N = 52$ or equivalently,

105 data points, we have

$$\cos 1000 t_j = \cos 40 t_j,$$
$$\sin 1000 t_j = -\sin 40 t_j,$$
$$\cos 589 t_j = \cos 35 t_j,$$
$$\sin 589 t_j = -\sin 35 t_j,$$
$$\cos 177 t_j = \cos 31 t_j,$$
$$\sin 177 t_j = -\sin 31 t_j,$$

where $t_j = j\pi/52$, $j = 0, 1, \ldots, 104$.

Thus, if we find the discrete Fourier series for $f(t)$ at these data points, viz.

$$f(t_j) = \frac{A_0}{2} + \sum_{r=1}^{51} (A_r \cos rt_j + B_r \sin rt_j) + \frac{A_{52}}{2} \cos 52 t_j,$$

we can say that at the data points

$$f(t) = \frac{A_0}{2} + \sum_{r=1}^{30} (A_r \cos rt + B_r \sin rt)$$
$$+ A_{31} \cos 177 t - B_{31} \sin 177 t$$
$$+ A_{35} \cos 589 t - B_{35} \sin 589 t$$
$$+ A_{40} \cos 1000 t - B_{40} \sin 1000 t,$$

within an error depending on the size of the Fourier coefficients of the slowly varying part of $f(t)$. We forego a discussion of the error analysis of this aliasing procedure, for which we refer to Snider, 1972. Instead we turn to the application of this process to the highly oscillatory stiff system.

4.3.4. Application to Highly Oscillatory Systems

We use Certaine's method which is discussed in Section 3.1.1. We see from that discussion that for the differential equation

$$y'(t) = -Dy(t) + g(y(t), t),$$

where y and g are m-vectors and D is an $m \times m$ constant matrix with at least one large eigenvalue, that Certaine's method is given by two utilizations of the following expression,

$$y_{n+1} = e^{-Dh} y_n + e^{-Dt_{n+1}} \int_{t_n}^{t_{n+1}} e^{D\tau} g_k(\tau) d\tau. \tag{4.3.4}$$

Here $g_k(t)$ is an interpolation polynomial of degree k which approximates $g(y, t)$ at the points $t_{n-k}, t_{n-k+1}, \ldots, t_n$.

In the oscillating case at hand, the polynomial g_k is replaced by a trigonometric polynomial. In this case as well, the integral in (4.3.4) may be explicitly evaluated. However, we will have an inefficient procedure unless we use aliasing. That is, we must know the higher frequencies in the problem (i.e., the large imaginary eigenvalues of D), and then we must alias these higher frequencies so that g_k is a trigonometric polynomial of low degree. For illustrative computations which employ this method, we refer once more to Snider and Fleming, 1974.

A criticism of this method arises in the case of a non-linear system. For in such a system, even though the frequencies are known to start with, we may find, among other nonlinear effects, the introduction of sum and difference frequencies into the solution as it develops. Of course the determination of N, depending on L and the $R_j, j = 1, \ldots, p$, requires a computation also.

4.4. FITTING IN THE CASE OF PARTIAL DIFFERENTIAL EQUATIONS

Partial differential equations of evolutionary type along with their numerical treatment are subject to being ill-conditioned. In some cases this ill-conditioning resembles the difficulties associated with stiff ordinary differential equations. The remedy of exponential fitting for the latter has a counterpart for partial differential equations, and we review this counterpart here. As we might expect in the partial differential equations case, the idea of exponential fitting is susceptible to a much wider scope of possibilities and results than in the ordinary differential equations case.

To motivate our discussion, we begin with a review of a model problem and an elementary error analysis.

4.4.1. *The Problem Treated*

Let D be the domain of points, $D = \{(x, t) \,|\, t \in (0, T], |x| < \infty\}$ and consider the initial value problem

$$
\begin{aligned}
u_t &= \lambda u_x, & (x, t) \in D, \quad t \neq 0, \\
u(x, 0) &= f(x), & t = 0.
\end{aligned}
\tag{4.4.1}
$$

Here λ is a scalar, and u and f are real valued scalar functions. This model problem has the solution

$$
u(x, t) = f(x + \lambda t).
\tag{4.4.2}
$$

In the half plane, $t \geq 0$, we set down a mesh, $\mathcal{M}$, with increments Δt and Δx, i.e., $\mathcal{M} = \{(x_j, t_n) = (j\Delta x, n\Delta t), \ j = 0, \pm 1, \ldots, n = 0, 1, \ldots\}$. We may suppose without loss of generality that $\Delta t = \Delta x = h$.

Let $u_n \equiv u_n(x) \equiv u(x, nh)$. Then

$$u_{n+1} = S\left(h\frac{\partial}{\partial x}\right)u_n, \quad n = 0, 1, \ldots,$$

$$S(z) = e^{\lambda z},$$

as (4.4.2) shows.

As a numerical approximation to u_n, we take $v_n \equiv v_n(x)$, where

$$v_{n+1} = \sum_{|j| \leq l} a_j H^j v_n, \quad n = 0, 1, \ldots,$$

$$v_0 = f(x). \tag{4.4.3}$$

Here $l \geq 0$ is an integer, and H is the shift operator, $Hf(x) = f(x + h)$. (4.4.3) is commonly called a *two level explicit difference scheme*. If $|a_l| + |a_{-l}| \neq 0$, we say that this scheme has *width l*. We write (4.4.2) as

$$v_{n+1} = Kv_n, \tag{4.4.4}$$

where K is the amplification operator of the scheme.

If the powers $\| K \|^j, j = 1, 2, \ldots$ are bounded, then the numerical scheme is stable, and we may obtain the following bound for the global error, $e_n = v_n - u_n$.

$$\| e_n \| \leq \text{const} \times n \times \max_{0 \leq p \leq n} \| Tu_p \|. \tag{4.4.5}$$

Here $T = K - S$ is the truncation operator, and we are using $\| \cdot \|$ to denote the norm in $L^2[-\infty, \infty]$.

Using Taylor's theorem and the following consistency relations for the difference scheme (4.4.3):

$$1 - \sum_{j \leq l} a_j = 0,$$

$$\lambda - \sum_{j \leq l} ja_j = 0, \tag{4.4.6}$$

(4.4.5) becomes

$$\| e_n \| \leq \text{const} \times nh^2 \times \max_{0 \leq p \leq n} \| u_p''(\eta) \|, \quad \eta \in (x - lh, x + lh). \tag{4.4.7}$$

If u_p'' exists and is bounded by a constant M uniformly in the domain D, the bound (4.4.7) becomes

$$\| e_n \| \leq \text{const} \times Mh, \tag{4.4.8}$$

provided that $nh \leq T$.

As the data $f(x)$ or the solution $u_p(x)$ becomes less smooth, the bound (4.4.8) becomes less satisfactory, and convergence of the pointwise error to zero with h becomes slower and slower. Indeed, when the data or solution becomes discontinuous, there is no bound M at all, and the convergence of the pointwise error is a delicate question. This difficulty in turn is reflected in an inadequate state of affairs in actual computations for such problems. The problems are ill-conditioned. Indeed, as the data becomes less smooth, the absolute values of its Fourier transform at larger frequencies tend to grow. Since the spectrum of $\lambda \partial / \partial x$ is continuous, we see then that as the operator $S(\lambda \partial / \partial x)$ develops the solution, it receives increasing input at higher frequencies as the data degrades (i.e., as the data lessens in smoothness).

We see then that the situation is quite analogous to the case of stiff systems of ordinary differential equations.

What we will do is to return to the bound (4.4.5) for $\| e_n \|$, and make $\| Tu_p \|$ as small as possible. That is, we will minimize $\| Tu_p \|$ over the set of real coefficient vectors $a \equiv (a_{-l}, \dots, a_l)$. An alternative approach would be to minimize the $\max_{u_p} \| Tu_p \|$, a procedure which resembles the minimax fitting discussed in Section 4.1.2. We will not discuss this possibility here, but refer to Micchelli and Miranker, 1973 and 1974 for details concerning it. Instead we consider a set of special cases in which we replace this maximization over u_p by an appropriate choice of u_p itself. The principle being that if we seek to derive a numerical method with desirable properties relative to a given type of problem (or data), we cause the properties which are wanted to be taken on by constraining the minimization or fixing the weight function u_p appropriately. We will henceforth drop this subscript p.

4.4.2. The Minimization Problem

To formulate the minimization problem to be considered, we introduce the Fourier transform $\hat{f}$ of f, where

$$\hat{f} = \hat{f}(\omega) = \frac{1}{\sqrt{2\pi}} \int_{-\infty}^{\infty} e^{-i\omega x} f(x) \mathrm{d}x.$$

Then the minimization problem becomes

$$\min_a \| Tu \| = \min_a \| (\hat{K} - \hat{S})\hat{u} \|. \qquad (4.4.9)$$

Here

$$\hat{K}(z) = \sum_{|j| \le l} a_j e^{ijz},$$
$$\hat{S}(z) = e^{i\lambda z}.$$

Then the function to be minimized is

$$J = \| (\hat{K} - \hat{S})\hat{u} \|^2 = \int_{-\infty}^{\infty} |\hat{K}(h\omega) - \hat{S}(h\omega)|^2 |\hat{u}(\omega)|^2 d\omega. \qquad (4.4.10)$$

We introduce some terminology in the following definition and the ensuing remark.

DEFINITION 4.4.1. We call the schemes (4.4.3), which use the vector of coefficients, a, determined by the minimization problem (4.4.9), schemes with best possible (local-) truncation error, or simply *best possible schemes*.

REMARK 4.4.2. Schemes for which the $2l + 1$ degrees of freedom represented by a are chosen so as to achieve the relation

$$\hat{K}(h\omega) = \hat{S}(h\omega) + O((h\omega)^p), \qquad (4.4.11)$$

with $p = 2l$ are the classical schemes. These are schemes of maximal order or of maximal (local-) accuracy. They have been named the *most accurate schemes* (see Strang, 1962).

For an integer $p \le 2l$, the relation (4.4.11) is equivalent to the following $p + 1$ moment conditions.

$$\sum_{|j| \le l} j^r a_j = \lambda^r, \qquad r = 0, 1, \ldots, p, \qquad p \le 2l.$$

4.4.3. Highly Oscillatory Data

Derivation of the Quadratic Form
For problems with highly oscillatory data, a good choice of $u(x)$ is one such that

$$|\hat{u}(\omega)|^2 = \begin{cases} 1, & |\omega| < c/h, \\ 0, & |\omega| > c/h, \end{cases}$$

for c, some constant. In this case, we denote J (see (4.4.10)) by J_c. Evidently,

$$J_c = \frac{2h}{c} \int_{-c/h}^{c/h} |\hat{K}(h\omega) - \hat{S}(h\omega)|^2 d\omega).$$

For $c = \pi/\lambda$, we find

$$J_{\pi/\lambda} = 1 + \sum_{|j| \leq l} a_j^2 - \frac{2\lambda}{\pi} a_j \frac{\sin(j\pi/\lambda)}{\lambda - j}$$

$$+ \frac{2\lambda}{\pi} \sum_{k=1}^{2l} \left(\sum_{j_1 - j_2 = k} a_{j_1} a_{j_2} \right) \frac{\sin(k\pi/\lambda)}{k}.$$

For $c = \pi$, we find

$$J_\pi = 1 + \sum_{|j| \leq l} a_j^2 - 2 \frac{\sin \lambda \pi}{\pi} \sum_{|j| \leq l} (-1)^j \frac{a_j}{\lambda}. \tag{4.4.12}$$

For $c = p\pi$, p an integer, we find

$$J_{p\pi} = 1 + \sum_{|j| \leq l} a_j^2.$$

Consistent Formulas

Let us minimize J_π with respect to a and subject to the *constraints of consistency* (4.4.6). We may expect the resulting finite difference scheme to be good uniformly over all frequencies. Using the method of Lagrange multipliers, we find the minimizing value of a_j, $|j| \leq l$ to be

$$a_j = \frac{1}{2l + 1} \left[1 - \rho \sum_{|k| \leq l} \frac{(-1)^k}{\lambda - k} \right]$$

$$+ \frac{1}{2S_2} \left[\lambda - \rho \sum_{|k| \leq l} \frac{(-1)^k k}{\lambda - k} \right] + \frac{(-1)^j}{\lambda - j} \rho, \tag{4.4.13}$$

$$\rho = \frac{\sin \lambda \pi}{\pi}, \quad S_2 = \tfrac{1}{6} l(l+1)(2l+1).$$

If in (4.4.13) we set $\lambda = m$, an integer, we get

$$a_j = \delta_{jm}.$$

In this case, the difference scheme propagates information precisely along the characteristic of the partial differential equation, i.e., the numerical solution is exact.

In the case $l = 1$, (4.4.13) becomes

$$a_0 = \frac{1}{3}\left[1 + \frac{\lambda^2 + 1}{\lambda(\lambda^2 - 1)}\rho\right] + \frac{\rho}{\lambda},$$

$$a_{\pm 1} = \frac{1}{3}\left[1 + \frac{\lambda^2 + 1}{\lambda(\lambda^2 - 1)}\rho\right] \pm \frac{1}{2}\left[\lambda + \frac{2}{\lambda^2 - 1}\rho\right] - \frac{1}{\lambda \mp 1}.$$

We similarly find that the minimum of $J_{p\pi}$ is taken on when

$$a_j = \frac{1}{2l + 1} + \frac{j\lambda}{2S_2}.$$

This scheme is always stable. To see this, note that

$$\min_{|j| \le l} a_j = a_{-1} = \frac{p(l + 1) - 3}{p(l + 1)(2l + 1)} > 0,$$

since $p \ge 2$ (consistency), $l \ge 1$, and appeal to the following lemma whose proof is straightforward.

LEMMA 4.4.3. *Difference schemes of the type (4.4.3) for which* $\sum_{j \le l} a_j = 1$ *and* $a_j \ge 0, j = 0, \pm 1, \ldots, \pm l$ *are stable.*

Consistent Formulas which are Fitted at High Frequency
If the data has large frequency components, the constraints

$$\hat{T}(z)\big|_{z = \pm c/h} = 0$$

suggest themselves. Setting $c = p$, let us minimize $J_{p\pi}$ subject to these constraints along with the constraints of consistency, *i.e.*, subject to the following for constants.

$$\hat{T}(0) = \hat{T}'(0) = \hat{T}(p\pi/h) = \hat{T}(-p\pi/h) = 0.$$

The minimum occurs at

$$a_j = \begin{cases} \dfrac{1}{2l + 1} + \dfrac{j}{2pS_2}, & p \text{ even,} \\[2ex] [1 - (-1)^j]\left[\dfrac{(-1)^l}{l(1 + 2l)} + \dfrac{1}{2l}\right] + \dfrac{j}{2pS_2}, & p \text{ odd.} \end{cases} \qquad (4.4.14)$$

In the case of even p, we deduce from (4.4.14) that

$$\min_{|j| \le l} a_j = a_{-l} = \frac{p(l+1) - 3}{p(l+1)(2l+1)} > 0,$$

since $p, l \ge 2$. Thus in the case of even p, the schemes given by (4.4.14) are always stable.

4.4.4. Systems

Derivation of the Quadratic Form
The approach to the determination of difference schemes which we are discussing may be carried over directly to the case of systems of first order partial differential equations.

Let u and v be q-vectors and let A be a $q \times q$ matrix. We consider the initial value problem for

$$u_t = A u_x, \qquad (x, t) \in D, \qquad t \ne 0. \tag{4.4.15}$$

The difference scheme is

$$v_{n \ne 1} = \sum_{j \le l} B_j H^j v_n, \tag{4.4.16}$$

where the $B_j, |j| \le l$, are $q \times q$ matrices.

Proceeding as before by taking Fourier transforms of (4.4.15) and (4.4.16), we are led to the problem of minimizing the following functional

$$J = \| (\hat{K}(h\omega) - \hat{S}(h\omega)) \hat{u}(\omega) \|^2.$$

Here $\hat{K}(z)$ and $\hat{S}(z)$ are the $q \times q$ matrices given by

$$\hat{K}(z) = \sum_{|j| \le l} B_j e^{ijz}$$

$$\hat{S}(z) = e^{izA},$$

respectively.

For the weight function, we choose $\hat{u}(\omega)$ to be

$$\hat{u}(\omega) = \eta \times \begin{cases} 1, & |\omega| \le \pi/h, \\ 0, & \text{otherwise,} \end{cases}$$

Here η is the q-vector all of whose components are unity. This choice of $\hat{u}(\omega)$ makes J correspond to the functional J_π in (4.4.12).

Now using $(\cdot, \cdot)$ to denote the inner product in Euclidean q-space, we may rewrite J as follows.

$$J = \int_{-\pi}^{\pi} [(\hat{S}(z) - \hat{K}(z))\eta, (\hat{S}(z) - \hat{K}(z))\eta] dz. \tag{4.4.17}$$

Now suppose that A is a symmetric matrix with eigenvalues, λ_i, $i = 1, \ldots, q$. Let U be the unitary matrix which diagonalizes A, viz.

$$UAU^{-1} = \Lambda,$$

where Λ is the diagonal matrix whose iith entry is λ_i, $i = 1, \ldots, q$. Let $UB_j U^{-1} = C_j$, and let $U\eta = \mu$. Then (4.4.17) becomes

$$J = \int_{-\pi}^{\pi} \left((e^{iAz} - \sum_{|j| \le l} C_j e^{ijz})\mu, \ (e^{iAz} - \sum_{|j| \le l} C_j e^{ijz})\mu \right) dz.$$

Now let $C_j = (c_{mn}^j)$, $m, n = 1, \ldots, q$, and let $\mu = (\mu_1, \ldots, \mu_q)^T$. Also let

$$\gamma_m^j = \sum_{n=1}^{q} c_{mn}^j \mu_n.$$

Then J becomes

$$J = \sum_{m=1}^{q} \left[\mu_m^2 + \sum_{|j| \le l} (\gamma_m^j)^2 - 2 \frac{\sin \lambda_m \pi}{\pi} \sum_{|j| \le l} (-1)^j \frac{\gamma_m^j \mu_m}{\lambda - j} \right].$$

The constraints of consistency are

$$\sum_{|j| \le l} j^k B_j = A^k, \qquad k = 0, 1,$$

or equivalently

$$\sum_{|j| \le l} j^k C_j = \Lambda^k, \qquad k = 0, 1.$$

An Example

Let us consider the case corresponding to the wave equation. Here the dimension $q = 2$ and $A = c \begin{bmatrix} 0 & 0 \\ -1 & 1 \end{bmatrix}$, where c is the sound speed. Then we find that

$$U = \frac{1}{\sqrt{2}} \begin{bmatrix} 1 & -1 \\ 1 & 1 \end{bmatrix}, \lambda_1 = -\lambda_2 = c \quad \text{and} \quad \gamma_m^j = \sqrt{2} \, c_{m1}^j.$$

J becomes

$$J = 2\left[1 + \sum_{j=-1}^{1}\left[(c_{11}^{j})^{2} + (c_{21}^{j})^{2}\right] - 2\frac{\sin c\pi}{\pi}\sum_{j=-1}^{1}(-1)^{j}\frac{c_{11}^{j}}{c-j}\right].$$

The solution to the constrained minimization problem is found to be

$$B_{1} = \frac{1}{2}\gamma_{-1}M_{1} + \frac{c}{4}M_{2} + \frac{1}{4}M_{3},$$

$$B_{0} = \frac{1}{2}\gamma_{0}M_{1} - \frac{1}{4}M_{3}, \tag{4.4.18}$$

$$B_{1} = \frac{1}{2}\gamma_{1}M_{1} - \frac{c}{4}M_{2} - \frac{1}{4}M_{3}.$$

Here

$$M_{1} = \begin{bmatrix} 1 & 1 \\ 1 & 1 \end{bmatrix}, \quad M_{2} = \begin{bmatrix} 1 & -1 \\ -1 & 1 \end{bmatrix}, \quad M_{3} = \begin{bmatrix} \alpha - \beta & -\alpha + \beta \\ \alpha + \beta & -\alpha - \beta \end{bmatrix}.$$

α and β are arbitrary parameters, and

$$\gamma_{-1} = \frac{1}{3} - \frac{c}{2} - \frac{1}{6}\frac{\rho}{c+1}\frac{2c^{2}-1}{c(c-1)},$$

$$\gamma_{0} = \frac{1}{3} + \frac{1}{2}\frac{\rho}{c}\left(1 + \frac{1}{3}\frac{1+c^{2}}{c^{2}-1}\right),$$

$$\gamma_{1} = \frac{1}{3} + \frac{c}{2}\frac{\rho}{6(c-1)}\frac{2c^{2}-1}{c(c+1)},$$

with $\rho = 2 (\sin c\pi)/\pi$.

α and β may be chosen so that the resulting difference scheme is more like the usual scalar scheme. This is accomplished by demanding that

$$B_{k} = P_{k}(A), \qquad k = 0, \pm 1,$$

where the $P_{k}(A)$ are polynomials in A. We find that $\alpha = \beta = 0$ and that

$$P_{-1}(A) = \left(\frac{1}{2}\gamma_{-1} + \frac{c}{4}\right)I + \left(\frac{1}{4} - \frac{\gamma_{-1}}{2c}\right)A,$$

$$P_{0}(A) = \frac{1}{2}\gamma_{0}I + \frac{1}{2c}\gamma_{0}A,$$

$$P_{1}(A) = \left(\frac{1}{2}\gamma_{1} - \frac{c}{4}\right)I - \left(\frac{1}{4} + \frac{\gamma_{1}}{2c}\right)A.$$

Here I is the 2×2 identity matrix.

4.4.5. Discontinuous Data

The Scalar Case
We return to the scalar case and consider the problem of optimizing
the difference scheme when the data is discontinuous. Thus we are interes-
ted in minimizing $\|Tu\|$ when

$$u(x) = \begin{cases} 1, & x \geq 0, \\ 0, & x < 0. \end{cases}$$

In this case,

$$\sqrt{2\pi}\hat{u}(\omega) = \lim_{d \to \infty} \int_{-d}^{d} e^{-\omega x} u_n(x) dx = \lim_{d \to \infty} \frac{i}{\omega}[e^{-\omega d} - 1].$$

With this choice of $\hat{u}(\omega)$ and the associated weight function $|\hat{u}(\omega)|^2$, we
denote the corresponding value of $\|Tu\|^2$ by J_D. Then

$$2\pi J_D = \lim_{d \to \infty} \int_{-\infty}^{\infty} |T(h\omega)|^2 \left|\frac{e^{-i\omega d} - 1}{\omega}\right|^2 d\omega$$

$$= 2\int_{-\infty}^{\infty} |T(h\omega)|^2 \frac{d\omega}{\omega^2} - 2 \lim_{d \to \infty} \int_{-\infty}^{\infty} |T(h\omega)|^2 \frac{\cos \omega d}{\omega^2} d\omega.$$

Both integrals exist if $T(0) = 0$, a property implied by the consistency of the
scheme, which we will always require. The last integral here tends to zero
when $d \to \infty$, as an integration by parts shows.

A straightforward calculation now gives the value of J_D, which is

$$J_D = 2l\left[\sum_{|j| \leq l} |\lambda - j| a_j - \sum_{k=1}^{2l} k\left(\sum_{j_1 - j_2 = k} a_{j_1} a_{j_2}\right)\right].$$

We now minimize J_D over the vectors a and subject to the constraints of
consistency, (4.4.6). For the minimizing a, we find that

$$a_j = \begin{cases} \lambda + 1 - j, & j - 1 \leq \lambda \leq j, \\ -\lambda + 1 - j, & j \leq \lambda \leq j + 1, \\ 0, & \text{otherwise}, \quad |j| \leq l. \end{cases} \tag{4.4.19}$$

These $a_j(\lambda)$ are the translates of the cardinal spline of order unity. From
(4.4.19) we see that only those coefficients corresponding to mesh points
which immediately surround the characteristic of the differential equation
(4.4.1), which passes through the forward time point $(x, (n+1)\Delta t)$, are

non-zero. Notice also that the a_j given in (4.4.19) are non-negative. Thus this most accurate scheme is always stable (see Lemma 4.4.3).

Systems
The procedure discussed in Section 4.4.4 for a system may be carried over to the case of the discontinuous data at hand. The details are quite similar, and we merely display the following analogue of (4.4.18).

$$B_{-1} = \frac{1}{4}\begin{bmatrix} 1 & 1 \\ 1 & 1 \end{bmatrix} + \frac{c}{4}\begin{bmatrix} 1 & 1 \\ 1 & 1 \end{bmatrix} - \frac{1}{4}\begin{bmatrix} \alpha & \alpha \\ \beta & \beta \end{bmatrix},$$

$$B_0 = \frac{1}{2}\begin{bmatrix} 1 & -1 \\ -1 & 1 \end{bmatrix} + \frac{c}{2}\begin{bmatrix} -1 & 1 \\ 1 & -1 \end{bmatrix} + \frac{1}{2}\begin{bmatrix} \alpha & \alpha \\ \beta & \beta \end{bmatrix},$$

$$B_1 = \frac{1}{4}\begin{bmatrix} 1 & 1 \\ 1 & 1 \end{bmatrix} + \frac{c}{4}\begin{bmatrix} 1 & -3 \\ -3 & 1 \end{bmatrix} - \frac{1}{2}\begin{bmatrix} \alpha & \alpha \\ \beta & \beta \end{bmatrix}.$$

Casting B_j into the form $P_j = a_j I + b_j A, j = 0, \mp 1$ where a_j and b_j are scalars and I is the 2×2 identity matrix, we find that $\alpha = \beta$ and

$$P_{-1}(A) = \frac{1+c-\alpha}{4} I - \frac{1+c-\alpha}{4c} A,$$

$$P_0(A) = \frac{1-c+\alpha}{2} I + \frac{1-c-\alpha}{2c} A, \qquad (4.4.20)$$

$$P_1(A) = \frac{1+c-\alpha}{4} I - \frac{1-3c-\alpha}{2c} A.$$

A simple and interesting special case of (4.4.20) corresponds to setting $\alpha = c$.

4.4.6. *Computational Experiments*

In this section, we present the results of calculations based on some of the difference schemes derived above. We have chosen three classes of schemes for the calculations. The first is the class of most accurate schemes. This class of schemes is given by

$$a_j = \prod_{\substack{|k| \le l \\ k \neq j}} \frac{\lambda - k}{j - k}, \qquad j = -l, \ldots, l.$$

See Remark (4.4.2.). This class of schemes is viewed as the test case against which the results of any other schemes are to be compared. We expect that

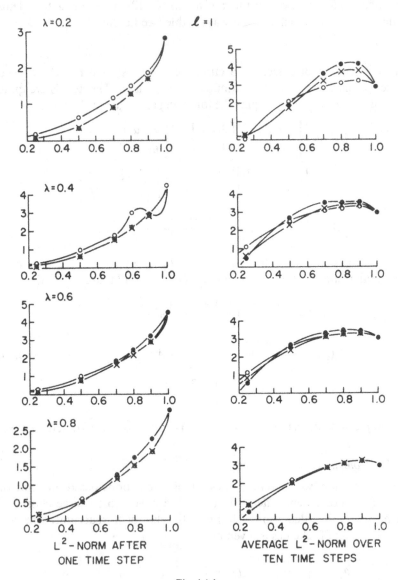

Fig. 4.4-1.

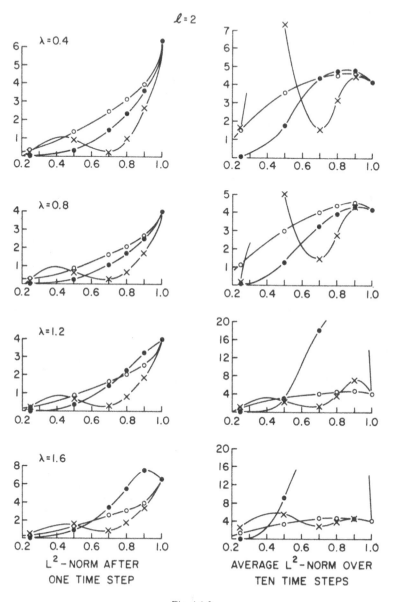

Fig. 4.4-2.

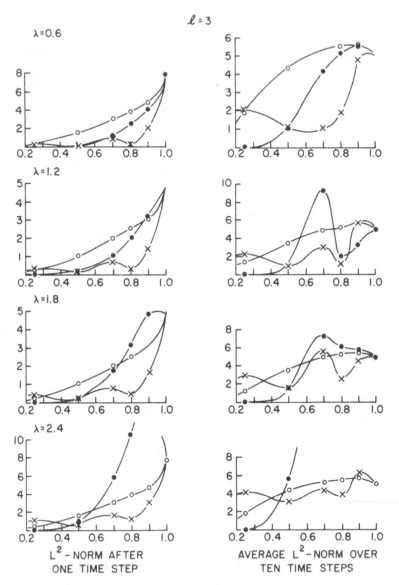

$\ell = 3$

L^2 - NORM AFTER
ONE TIME STEP

AVERAGE L^2-NORM OVER
TEN TIME STEPS

Fig. 4.4-3.

for data with low frequencies only, i.e., smooth or non-stiff problems, this class is probably as good a set of methods as possible.

The second class is those schemes which are best possible for data which is a step function, i.e., the scheme (4.4.19). The third class is the scheme (4.4.13). This is the class of best possible consistent methods which weighs all frequencies equally.

We solve the initial value problem

$$u_t = \lambda u_x, \qquad -\infty < x < \infty, \quad t > 0,$$

$$u(x, 0) = \sin k\pi x, \qquad -\infty < x < \infty.$$

The mesh increments Δt and Δx are taken without loss of generality to be unity. Then we see that the frequency parameter k of the data can have a sensible maximum of unity since this is the highest frequency which can exist on the mesh. We solve this initial value problem numerically with each of the three classes of schemes discussed above. k varies between 0.25 and 1.0 while λ varies between zero and l, the *Courant number* for the problem. (The smooth problems correspond to the smaller values of k, while the stiffer problems correspond to values of k near unity.)

We illustrate the results of these computations by means of graphs in Figures 4.4–1, 4.4–2 and 4.4–3. One set of graphs is a plot of the discrete L^2 norm of the pointwise error after one integration step versus frequency k. The error is computed over a range of x, $10(2l + 1)\Delta x$ wide, i.e., for the points $x = -10l, -10l + 1, \ldots, 10l$. (Recall that $\Delta x = 1$.) In the second set of graphs, we plot the average of the discrete L^2 norm of the pointwise error, as just described; the average taken over the first ten time steps.

In the graphs, the three classes of schemes are distinguished by plotting the first class with dots, the second with circles and third with crosses. The abscissa is k and the ordinates are unitless.

The results clearly show that at low frequencies (i.e., smooth problems) the class of most accurate schemes is best. At higher frequencies (i.e., stiffer problems) this class is superseded by both the remaining two classes, although, the third class, i.e., (4.4.13) is usually the best of all at higher frequencies. The plots for ten time steps give a similar picture, but show that the second class has a better average performance over all frequencies.

Finally, note that schemes in the first class, the class of most accurate schemes, are unstable for $\lambda > 1$ and so are not competitive in this range in any event. Some of the graphs show this instability clearly. For further details concerning fitting in the case of partial differential equations, we refer to Miranker, 1971.

Chapter 5

Methods of Boundary Layer Type

Summary

We begin our study of numerical methods for singular perturbation problems and the connection of these methods to the numerical treatment of stiff differential equations by considering the generic initial value problem for a singularly perturbed system of differential equations written in the following form.

$$\frac{dx}{dt} = f(t, x, y, \varepsilon), \qquad x(0) = \xi,$$

$$\varepsilon \frac{dy}{dt} = g(t, x, \varepsilon), \qquad y(0) = \eta,$$

(5.1)

where $x(t), f \in R^m$ and $y(t), g \in R^n$. We assume that f and g depend regularly on ε and that $g(t, x, y, 0) \neq 0$.

In Section 1.3, we noted that this class of systems is stiff by observing that in the case that $f = y$ and $g = x + y$, the eigenvalues of the system are $\varepsilon^{-1} + O(1)$ and $-1 + O(\varepsilon)$. In a sense, the smaller ε, the stiffer the system. Thus the large collection of analytic methods, commonly called boundary layer methods, used to characterize solutions of singularly perturbed systems, could be exploited to generate numerical methods for stiff systems. Since the approximations produced by these analytic methods improve with decreasing ε, we may expect that the corresponding numerical methods will likewise improve with increasing stiffness in the system. Of course the case is typically just the opposite in our previous discussions.

Since the solution of the system (5.1) is described by the so-called boundary layer formalism, we will refer to numerical methods developed according to this idea as *numerical methods of boundary layer type*.

We begin in Section 5.1 with a description of the boundary layer formalism and a description of the boundary layer numerical method. Then in Section 5.2 we discuss the so-called ε-independent method which connects the singular perturbation technique rather directly to a larger class of stiff problems. This is followed by the results of computational experiments with this method. Finally, in Section 5.3 we present an extrapolation method which allows the stiff system to be dealt with by treatment of a pair of associated non stiff or relaxed equations.

5.1. THE BOUNDARY LAYER NUMERICAL METHOD

5.1.1. *The Boundary Layer Formalism*

We begin with a review of the formalism of boundary layers. Although the formalism is simply outlined here, a derivation exhibiting the rationale upon which it is built is given below in Section 5.2 in the context of the ε-independent method. The solutions $x(t)$ and $y(t)$ of (5.1) have expansions of the type

$$x(t) \sim \sum_{r=0}^{\infty} x_r(t)\frac{\varepsilon^r}{r!} + \sum_{r=0}^{\infty} X_r(\tau)\frac{\varepsilon^r}{r!}, \qquad (5.1.1)$$

$$y(t) \sim \sum_{r=0}^{\infty} y_r(t)\frac{\varepsilon^r}{r!} + \sum_{r=0}^{\infty} Y_r(\tau)\frac{\varepsilon^r}{r!}, \qquad (5.1.2)$$

where

$$\tau = t/\varepsilon. \qquad (5.1.3)$$

The symbol $\sim$ is used to denote the fact that the series in (5.1.1) and (5.1.2) are asymptotic expansions. The first and second sums in (5.1.1) and (5.1.2) are called the outer solution and the boundary layer, respectively.

Following well-known procedures (see Hoppensteadt, 1971 and Levin and Levinson, 1954), we find that the coefficients $\{x_r, y_r\}$ of the outer solution are determined from

$$\dot{x}_0 = f(t, x_0, y_0, 0), \qquad (5.1.4_0)$$
$$0 = g(t, x_0, y_0, 0),$$

$$\dot{x}_r = f_x(t, x_0, y_0, 0)x_r + f_y(t, x_0, y_0, 0)x_r + Q_r,$$
$$\dot{y}_{r-1} = g_x(t, x_0, y_0, 0)x_r + g_y(t, x_0, y_0, 0)y_r + R_r, \qquad (5.1.4_r)$$
$$r = 1, 2, \dots.$$

The dot represents d/dt. f_x denotes the $m \times m$ matrix whose ijth component is the derivative of the ith component of f with respect to the jth component of x. f_y, g_x and g_y are similarly defined. Q_r and R_r depend on $t, x_0, y_0, \ldots, x_{r-1}, y_{r-1}, r = 1, 2, \ldots$. In particular,

$$Q_1 = f_\varepsilon(t, x_0, y_0, 0),$$
$$R_1 = g_\varepsilon(t, x_0, y_0, 0). \tag{5.1.5}$$

The subscript ε denotes $\partial/\partial\varepsilon$.

Notice that for each $r = 0, 1, 2, \ldots$, the first equation in (5.1.4$_r$) represents a system of differential equations, while the second represents a system of finite equations.

Continuing to follow well known procedures, we find the following equations:

$$X_0' = 0,$$
$$Y_0 = g(0, x_0(0) + X_0, y_0(0) + Y_0, 0), \tag{5.1.6$_0$}$$

$$X_r' = p_r,$$
$$Y_r = g_x(0, x_0(0) + X_0, y_0(0) + Y_0, 0)X_r$$
$$\quad + g_x(0, x_0(0), y_0(0) + Y_0, 0)Y_r + q_r, \tag{5.1.6$_r$}$$
$$r = 1, 2, \ldots,$$

from which the coefficients $\{X_r, Y_r\}$ of the boundary layer are determined. The prime represents $d/d\tau$. p_r and q_r depend only on τ, $x_0(0)$, $y_0(0)$, ..., $x_{r-1}(0), y_{r-1}(0), X_0, Y_0, \ldots, X_{r-1}, Y_{r-1}, r = 1, 2, \ldots$. In particular,

$$p_1(\tau) = f(0, \xi, y_0(0) + Y_0, 0) - f(0, \xi, y_0(0), 0). \tag{5.1.7}$$

Supplementing the equations (5.1.4$_r$) and (5.1.6$_r$) for the x_r, y_r, X_r and Y_r is the set of initial conditions:

$$x_r(0) + X_r(0) = \xi\delta_{r0},$$
$$y_r(0) + Y_r(0) = \eta\delta_{r0}, \qquad r = 0, 1, \ldots, \tag{5.1.8}$$

where δ_{r0} is the Kronecker delta. Since there is one condition for each pair of variables, the determination of the expansion is still not complete. We require an additional condition for the undetermined initial values in (5.1.8).

We require that the X_r, Y_r be boundary layers; namely that

$$\lim_{\tau \to \infty} X_r(\tau) = \lim_{\tau \to \infty} Y_r(\tau) = 0. \tag{5.1.9}$$

Now the specification of the coefficients in the expansions is complete, and we determine them in ordered groups of four; $\{X_r, x_r, y_r, Y_r\}$, $r = 0$, $1, \dots$, as follows.

From (5.1.4), (5.1.6), (5.1.8) and (5.1.9), we have for $r = 0$

(a) $\quad X_0' = 0,$ $\qquad\qquad\qquad\qquad \lim_{\tau \to \infty} X_0 = 0,$

(b) $\quad \dot{x}_0 = f(t, x_0, y_0, 0),$ $\qquad\qquad x_0(0) = \xi,$

(c) $\quad 0 = g(t, x_0, y_0, 0),$ $\qquad\qquad\qquad\qquad\qquad$ (5.1.10)

(d) $\quad Y_0' = g(0, \xi, y_0(0) + Y_0, 0),$ $\quad Y_0(0) = \eta - y_0(0).$

(5.1.10a) has the solution $X_0 = 0$, and the succeeding equations uniquely determine x_0, y_0 and Y_0. The condition (5.1.9) for Y_0 is satisfied if the eigenvalues of g_y, denoted $\lambda(g_y)$, satisfy

$$\lambda(g_y) < 0. \qquad\qquad\qquad\qquad\qquad\qquad (5.1.11)$$

The condition (5.1.11) characterizes the class of stiff systems to which the methods which we are now discussing are designed to be applied. We henceforth assume that (5.1.11) holds.

Similarly, for $r = 1$, we have

(a) $\quad X_1' = p_1(\tau),$ $\qquad\qquad\qquad\quad \lim_{\tau \to \infty} X_1(\tau) = 0,$

(b) $\quad \dot{x}_1 = f_x x_1 + f_y y_1 + f_\varepsilon,$ $\qquad x_1(0) = -X_1(0),$

(c) $\quad \dot{y}_0 = g_x x_1 + g_y y_1 + g_\varepsilon,$ $\qquad\qquad\qquad\qquad$ (5.1.12)

(d) $\quad Y_1' = g_x X_1 + g_j Y_1 + q_1,$ $\qquad Y_1(0) = -y_1(0).$

To solve (5.1.12), we proceed as follows. From (5.1.12a) we get

$$X_1(0) = -\int_0^\infty p_1(\sigma)d\sigma.$$

This and (5.1.8) determine $x_1(0) = -X_1(0)$ so that (5.1.12b) and (5.1.12c) may be solved simultaneously for x_1 and y_1. Then (5.1.12d) may be solved for Y_1. This procedure may now be repeated for each $r = 2, 3, \dots$.

5.1.2. The Numerical Method

We describe a numerical method which consists of constructing the formal boundary layer expansion by solving the equations which specify its terms numerically.

Let $h > 0$ be a mesh increment. Let $z = (x, y)^T$ and $Z = (X, Y)^T$ be

$N = m + n$ dimensional vectors. Then from (5.1.1) and (5.1.2),

$$z(h) = z_0(h) + \varepsilon z_1(h) + Z_0(h/\varepsilon) + \varepsilon Z_1(h/\varepsilon) + O(\varepsilon^2).$$

Since the equations are stiff, we are interested in the case

$$h \gg \varepsilon.$$

This and condition (5.1.9) imply that $Z_0(h/\varepsilon)$ and $Z_1(h/\varepsilon)$ will be near zero. In fact these terms will in general be exponentially small in h/ε. Thus we approximate $z(h)$ by $z_0(h) + \varepsilon z_1(h)$, the approximation being $O(\varepsilon^2)$ (i.e., it improves with increasing stiffness). The numerical method consists of calculating $z_0(h)$ and $z_1(h)$. We must still compute Z_0 in order to obtain the initial condition $x_1(0)$ required for the determination of $z_1(h)$. (Of course more terms in the expansion may be calculated if they are wanted.)

The numerical method consists of the following steps (i)–(iv):

(i) Solve

(a) $\dot{x}_0 = f(t, x_0, y_0, 0), \qquad x_0(0) = \xi,$

(b) $0 = g(t, x_0, y_0, 0)$

(5.1.13)

for $x_0(h)$, $y_0(0)$ and $y_0(h)$. The numerical method for solving (5.1.13a) should be of the self-starting type.

(ii) Having determined $y_0(0)$ in step (i), solve

$$Y_0' = g(0, \xi, y_0(0) + Y_0, 0), \qquad Y_0(0) = \eta - y_0(0)$$

for $Y_0(\tau)$, $\tau \geq 0$. This must be done for a net of τ-values, say $\{0, k, 2k, \dots, Mk\}$, so that

$$x_1(0) = -X_1(0) = \int_0^\infty p_1(\sigma) d\sigma$$

can be approximated to some prescribed degree of accuracy by a quadrature rule:

(iii)

$$\xi_1 = \sum_{j=0}^{M} a_j p_1(jk)$$

$$= \sum_{j=0}^{M} a_j [f(0, \xi, y_0(0) + Y_0(jk), 0) - f(0, \xi, y_0(0), 0)].$$

(iv) Having determined ξ_1, the approximation to $x_1(0)$, in step (iii), solve

(a) $\dot{x}_1 = f_x(t, x_0, y_0, 0) x_1 + f_y(t, x_0, y_0, 0) y_1$
              $+ f_\varepsilon(t, x_0, y_0, 0), \quad x_1(0) = \xi_1,$

(b) $y_1 = - g_y^{-1}(t, x_0, y_0, 0)[g_x(t, x_0, y_0, 0)x_1 - \dot{y}_0 + g_\varepsilon(t, x_0, y_0, 0)]$

for $x_1(h)$ and $y_1(h)$.

We make the following observations concerning this numerical method.

REMARK 5.1.1. Steps (i) and (iv) determine $z_0(h)$ and $z_1(h)$ respectively. Steps (ii) and (iii) deal with Z_0 and are used to determine the initial condition ξ_1 for x_1. The method seems to step across the rapidly varying modes (the boundary layers) as they change over the comparatively great interval $(0, h)$. This is not quite true, nor is it accomplished without cost. Steps (ii) and (iii) perform a mesh calculation with increment k in τ. Since $\tau = t/\varepsilon$, k will be $hO(\varepsilon)$. Thus, in order to calculate Z_0 and $x_1(0)$, a fine mesh calculation must be performed. The critique of this boundary layer method is:

(a) the parts or aspects of the given initial value problem upon which to perform the *fine mesh calculation* are a well-defined subpart of the original system.

(b) this fine calculation part may be performed with less precision than the coarse part (step (i)). To see that this is so, note that $z_1(h)$ depends on the fine part of the calculation through $x_1(0)$. Thus an error in determining the fine part leads to a proportional error in $z_1(h)$. But the approximation to the solution is $z_0(h) + \varepsilon z_1(h)$. Thus the effect of such an error is reduced in order by ε, a measure of the stiffness. Thus here again the stiffer the system, the more functional the method.

The following observation promises even more in this regard.

REMARK 5.1.2. In section 5.2, we will show how to by-pass this fine mesh calculation (see (5.2.22)).

5.1.3. *An Example*

We now consider an example for which the steps of the numerical method may be carried out analytically, i.e., to infinite arithmetic precision.

The example consists of the following initial value problem.

$$\dot{x} = y - x, \qquad\qquad x(0) = \xi,$$
$$\dot{y} = - 100y + 1, \qquad y(0) = \eta.$$

The exact solution of this problem is

$$x = \frac{1}{100} + \left[\xi + \frac{\eta - (1/100)}{99} - \frac{1}{100} \right] e^{-t} - \frac{\eta - (1/100)}{99} e^{-100t},$$

$$y = \frac{1}{100} + [\eta - (1/100)]e^{-100t}. \tag{5.1.14}$$

For the example, the steps of the numerical method are the following ones.
(i) Solve

$$\dot{x}_0 = y_0 - x_0, \qquad x_0(0) = \xi,$$
$$0 = y_0 \tag{5.1.15}$$

for $x_0(h)$, $y_0(0)$, and $y_0(h)$. We use Euler's method with increment h in t to solve (5.1.15). We find

$$x_0(h) = (1 - h)\xi,$$
$$y_0(0) = y_0(h) = 0. \tag{5.1.16}$$

(ii) and (iii) Solve

$$Y_0'(\tau) = - Y_0(\tau), \qquad Y_0(0) = \eta - y_0(0) = \eta \tag{5.1.17}$$

on the mesh $\tau_i = ik$, $i = 0, \dots, M$. Then evaluate

$$x_1(0) = \int_0^\infty Y_0(\sigma)d\sigma. \tag{5.1.18}$$

Using Euler's method with increment k in τ for (5.1.17) and using the rectangle rule for (5.1.18) with the upper limit of integration replaced by kM, we find

$$x_1(0) = \eta(1 - k^{M+1}).$$

(iv) Solve

$$\dot{x}_1 = 1 - x_1, \qquad x_1(0) = \eta(1 - k^{M+1}),$$
$$y_1 = 1.$$

Again using Euler's method with increment h, we find

$$x_1(h) = h + (1 - h)(1 - k^{M+1})\eta,$$
$$y_1(h) = 1. \tag{5.1.19}$$

Combining (5.1.16) and (5.1.19), we find

$$x(h) = (1 - h)\xi + \varepsilon(h + (1 - h)(1 - k^{M+1})\eta),$$
$$y(h) = \varepsilon. \tag{5.1.20}$$

Identifying ε with $1/100$, (5.1.20) becomes

$$x(h) = \frac{1}{100} + (1-h)\left(\frac{1}{100} + \xi + \frac{1}{100}(1-k^{M+1})\eta\right),$$

$$y(h) = \frac{1}{100},$$

which approximates (5.1.14) to the claimed accuracy.

5.2. THE ε-INDEPENDENT METHOD

A criticism of the boundary layer method which we have just discussed is that it depends on the stiff system being given in a form in which there is an identifiable small parameter which characterizes the system as one of singular perturbation type.

To deal with this criticism, we consider how boundary layer methods may be developed even though there is no identifiable small parameter. With this development, the boundary layer numerical method will be applicable to wider classes of stiff systems.

5.2.1. Derivation of the Method

We proceed by writing $k = (f, g)^T$, $z = (x, y)^T$, and $\zeta = (\xi, \eta)^T$. The initial value problem (5.1) is supposed given in the following form.

$$\dot{z} = k(t, z; \varepsilon), \qquad z(0) = \zeta. \tag{5.2.1}$$

Here and throughout this section, the parameter ε will appear explicitly in our development. In fact, ε is to be regarded as unidentifiable and as being displayed in some virtual sense. That is, we will suppose that the problem contains a small but unspecifiable parameter with respect to which certain properties are fulfilled. The derivation will make use of and display this parameter, in the virtual sense. What matters is that in the end result, the parameter is not needed and its utilization throughout is an artifact of the development. Of course the correctness (i.e., an error estimate) of the result depends on the correctness of the hypotheses concerning the unknown parameter. For a discussion of this latter point and other details, we refer to Miranker, 1973.

We solve the system (5.2.1) numerically along the mesh with increment h, proceeding as if the system were not stiff. In terms of the notation in Section 5.1.2, we start with m regarded as equal to the number of dimensions N, in z and with n equal to zero. The method then produces $z_0(h)$

by employing a standard self-starting numerical method. Then we compare $z_0(h)$ and ζ componentwise, i.e., we test the following inequality.

$$\frac{|z_{0,j}(h) - \zeta_j|}{1 + |\zeta_j|} > \theta, \qquad j = 1, \dots, N. \tag{5.2.2}$$

Here θ is a prescribed positive tolerance. If the tolerance is not exceeded by any component of $z_0(h)$, we accept the value of $z_0(h)$ produced. If the tolerance is exceeded by a set $\mathscr{J} = \mathscr{J}(j_1, \dots, j_n)$ of $n > 0$ components of $z_0(h)$, we reject the integration step and redo it as follows.

Set

$$\begin{aligned} x_i &= z_i, \\ \xi_i &= \zeta_i, \\ f_i &= k_i, \qquad i = 1, \dots, N, i \notin \mathscr{J}, \end{aligned} \tag{5.2.3}$$

and set

$$\begin{aligned} y_j &= z_j, \\ \eta_j &= \zeta_j, \\ g_j &= k_j, \qquad j = 1, \dots, N, \ j \in \mathscr{J}. \end{aligned} \tag{5.2.4}$$

Now the system has the form

$$\begin{aligned} \dot{x} &= f(t, z, \varepsilon), & x(0) &= \xi, \\ \dot{y} &= g(t, z, \varepsilon), & y(0) &= \eta. \end{aligned} \tag{5.2.5}$$

We stress that the parameter ε is still unidentifiable, and we make the following assumption.

ASSUMPTION 5.2.1. There exists a variable ε, so that $f(t, z, \varepsilon)$ and $g(t, z, \varepsilon)$ are analytic in ε in a neighborhood of $\varepsilon = 0$ except that $g(t, z, \varepsilon)$ has a simple pole at $\varepsilon = 0$. We also maintain the requirement $\lambda(g_y) < 0$ (see 5.1.1), assuring the boundary layer nature of the solution of the system.

We seek a solution of (5.2.5) in the form

$$x(t) = x_0(t) + \varepsilon x_1(t) + X_0(\tau) + \varepsilon X_1(\tau) + \dots, \tag{5.2.6}$$

$$y(t) = y_0(t) + \varepsilon y_1(t) + Y_0(\tau) + \varepsilon Y_1(\tau) + \dots. \tag{5.2.7}$$

For the outer solution, we have

$$\begin{aligned} \dot{x}_0 + \varepsilon \dot{x}_1 = f(t, x_0, y_0 ; \varepsilon) \\ + \varepsilon f_x(t, x_0, y_0 ; \varepsilon) x_1 + \varepsilon f_y(t, x_0, y_0, ; \varepsilon) y_1 + \dots, \end{aligned} \tag{5.2.8}$$

$$\dot{y}_0 + \varepsilon \dot{y}_1 = g(t, x_0, y_0; \varepsilon)$$
$$+ \varepsilon f_x(t, x_0, y_0; \varepsilon) x_1 + \varepsilon g_y(t, x_0, y_0,; \varepsilon) y_1 + \dots \quad (5.2.9)$$

By our assumption, the terms g, g_x and g_y have simple poles at $\varepsilon = 0$. Thus from (5.2.8) and (5.2.9), we deduce the following equations (5.2.10) and (5.2.11) for x_0, y_0, and for εx_1 and εy_1, (since ε is unidentifiable, we determine the products εx_1 and εy_1, and not x_1 and y_1, as before (see 5.1.12)), respectively.

(a) $\quad \dot{x}_0 = f(t, x_0, y_0; \varepsilon), \qquad x_0(0) = \xi, \qquad\qquad\qquad (5.2.10)$

(b) $\quad 0 = g(t, x_0, y_0; \varepsilon).$

Notice that consistent with its lack of identifiability, we do not set $\varepsilon = 0$. (Compare (5.2.10) with (5.1.4$_0$).)

For convenience we will hereafter suppress the arguments $(t, x_0, y_0; \varepsilon)$ of f and g. The equations for $\varepsilon \dot{x}_1$ and $\varepsilon \dot{y}_1$ are

$$\varepsilon \dot{x}_1 = f_x \varepsilon x_1 + f_y \varepsilon y_1,$$
$$\dot{y}_0 = g_x \varepsilon x_1 + g_y \varepsilon y_1. \qquad\qquad\qquad (5.2.11)$$

We solve the second equation here for εy_1 as follows.

$$\varepsilon y_1 = g_y^{-1}[\dot{y}_0 - g_x \varepsilon x_1] = g_y^{-1}[-g_y^{-1}(g_t + g_x f) - g_x \varepsilon x_1].$$
$$(5.2.12)$$

Here we replace $\dot{y}_0$ by its value obtained by differentiating (5.2.10b) with respect to t.

Combining (5.2.11) and (5.2.12) yields the following equations for determining εx_1 and εy_1, respectively.

$$\varepsilon \dot{x}_1 = (f_x - f_y g_y^{-1} g_x) \varepsilon x_1 - f_y g_y^{-2}(g_t + g_x f),$$
$$\varepsilon y_1 = g_y^{-1} g_x \varepsilon x_1 - g_y^{-2}(g_t + g_y f). \qquad (5.2.13)$$

Notice that ε is still unspecified, but the quantities εx_1 and εy_1 which are sought are, except for the initial condition $\varepsilon x_1(0)$, well-defined. Moreover, examining the right members of (5.2.13), we see by Assumption 5.2.1 that the large quantities g, g_x and g_y are neutralized, in the sense that they occur as quotients, one of the other.

To determine the initial condition, $\varepsilon x_1(0)$, we obtain an ε-independent determination of the boundary layers. Inserting (5.2.6) and (5.2.7) into

(5.2.5), we find

$$
\begin{aligned}
&\varepsilon x_0'(\varepsilon\tau) + \varepsilon^2 x_1'(\varepsilon\tau) + X_0'(\tau) + \varepsilon X_1'(\tau) + \ldots \\
&\quad = \varepsilon f(\varepsilon\tau, x_0(\varepsilon\tau) + \varepsilon x_1(\varepsilon\tau) + X_0(\tau) + \varepsilon X_1(\tau) \\
&\qquad + \ldots, y_0(\varepsilon\tau) + \ldots; \varepsilon), \\
&\varepsilon y_0'(\varepsilon\tau) + \varepsilon^2 y_1'(\varepsilon\tau) + Y_0'(\tau) + \varepsilon Y_1'(\tau) + \ldots \\
&\quad = \varepsilon g(\varepsilon\tau, x_0(\varepsilon\tau) + \varepsilon x_1(\varepsilon\tau) + X_0(\tau) + \varepsilon X_1(\tau) + \ldots, y_0(\varepsilon\tau) + \ldots; \varepsilon).
\end{aligned} \tag{5.2.14}
$$

Here and hereafter we use the prime to denote differentiation with respect to argument.

Using Assumption 5.2.1, we deduce the following equations for X_0, X_1 and Y_0 from (5.2.14).

First setting $\varepsilon = 0$ in (5.2.14), we obtain

$$
X_0'(\tau) = 0. \tag{5.2.15}
$$

As in Section 5.1, $X_0(0) = 0$, since $X_0(0) + x_0(0) = \xi$, so that $X_0(\tau) \equiv 0$. Next from (5.2.14), we deduce the following equations for X_1 and Y_0.

$$
X_1'(\tau) = f(0, \xi, y_0(0) + Y_0(\tau); \varepsilon) - f(0, \xi, y_0(0); \varepsilon), \tag{5.2.16}
$$

and

$$
Y_0'(\tau) = \varepsilon g(0, \xi, y_0(0) + Y_0(\tau); \varepsilon). \tag{5.2.17}
$$

We integrate (5.2.16) from zero to infinity, using the boundary layer property, $\lim_{\tau \to \infty} X(\tau) = 0$. Also using $x_1(0) + X_1(0) = 0$, we get

$$
\varepsilon x_1(0) = \varepsilon \int_0^\infty [f(0, \xi, y_0(0) + Y_0(\tau); \varepsilon) - f(0, \xi, y_0(0); \varepsilon)] \, d\tau. \tag{5.2.18}
$$

Now since $Y_0(\tau)$ vanishes exponentially fast as τ increases from zero, the bulk of the value of the integral in (5.2.18) comes from the neighborhood of $\tau = 0$. Thus we may expect a good approximation to the integral by replacing the integrand by an interpolant using data at $\tau = 0$. This data is first,

$$
Y_0(0) = \eta - y_0(0), \tag{5.2.19}
$$

from the initial condition, $y_0(0) + Y_0(0) = \eta$, while from (5.2.17) itself we have

$$
Y_0'(0) = \varepsilon g(0, \xi, \eta; \varepsilon). \tag{5.2.20}
$$

While we can obtain more data by differentiating (5.2.17), let us approximate (5.2.18) using just (5.2.19) and (5.2.20). The simplest approximation comes from replacing the integrand in (5.2.18) by its tangent at $\tau = 0$, and integrating this tangent from zero to its positive root. In this manner we obtain the following expression for approximating $\varepsilon x_1(0)$ from (5.2.18).

$$\varepsilon x_1(0) = \frac{1}{2} \frac{[f(0, \xi, \eta; \varepsilon) - f(0, \xi, y_0(0); \varepsilon)]^2}{f_y(0, \xi, \eta; \varepsilon) g(0, \xi, \eta; \varepsilon)}. \tag{5.2.21}$$

In (5.2.21) all arithmetic is componentwise except the matrix vector product $f_y g$ in the denominator. Notice that as far as magnitude with respect to ε is concerned, both sides of (5.2.20) are in agreement.

A second choice in approximating (5.2.18) is to use the data (5.2.19) and (5.2.20) to fit an exponential to the integrand, and then to integrate the exponential from zero to infinity. In this manner, we obtain the following expression for approximating $\varepsilon x_1(0)$ from (5.2.18).

$$\varepsilon x_1(0) = \frac{f(0, \xi, y_0(0); \varepsilon) - f(0, \xi, \eta; \varepsilon)}{f_y(0, \xi, \eta; \varepsilon) g(0, \xi, \eta; \varepsilon)}. \tag{5.2.22}$$

The arithmetic here is to be performed exactly as in the previous case.

With either (5.2.21) or (5.2.22), (5.2.13) determine εx_1 and εy_1 completely. We emphasize that while the derivation of (5.2.21) and (5.2.22) appears to depend upon and use the variable ε, it does so only in a virtual sense. Moreover the resulting approximating expressions for $\varepsilon x_1(0)$ in these two equations are independent of ε. (See the comment following (5.2.1).)

We now solve (5.2.10) for $y_0(0), y_0(h)$ and $x_0(h)$, by a numerical method as described earlier in Section 5.1.2. Then (5.2.13) and (5.2.21) or (5.2.22) are used to solve for $\varepsilon x_1(h)$ and $\varepsilon y_1(h)$ by a numerical method also described earlier. Finally we take

$$z(h) = \begin{pmatrix} x_0(h) + \varepsilon x_1(h) \\ y_0(h) + \varepsilon y_1(h) \end{pmatrix}. \tag{5.2.23}$$

We now repeat this procedure on the interval $(h, 2h)$. This time we start with the system already divided into a regular and singular part as in (5.2.5). We then make a tolerance test on $z(2h)$ compared with $z(h)$ analogous to (5.2.2). If the tolerance is not exceeded by any component of $z(2h)$, we accept the integration step. Otherwise we reject it and redivide the system according to the scheme described above (see (5.2.3)–(5.2.5)). We then redo this integration step. Once a component is placed into the singular part of the system, we do not remove it, even though its solution settles down and passes the tolerance test. Thus the flow of components of z from x status to y status is unidirectional. If this policy is not followed, the component in question usually regenerates a stiff mode (becomes unstable) at once and it is then pushed back into the singular part anyway. This aspect of the ε-independent numerical method concerning the tolerance test is an algorithmic aspect and should be adjusted to the particular problem being considered. It is likely that for nonlinear systems where the stiffness comes and goes as the solution evolves, a two-directional flow components of z between the regular and singular parts may be called for.

5.2.2. Computational Experiments

In this section, we give the results of calculations performed on two different stiff systems using the numerical method of Section 5.2.1 and two comparison methods.

For the numerical treatment of each of the equations of the boundary layer numerical method, we choose a simple numerical procedure. Thus for solving the differential equation (5.2.10a), we use the modified Euler method which for the equation $\dot{y} = f(t, y)$ is

$$y(t + h) = y(t) + \frac{1}{2}h[f(t, y(t)) + f(t + h, y(t) + hf(t, y))].$$

For solving the finite equation (5.2.10b), we use one iteration of Newton's method with initial guess η when solving for $y_0(0)$ and with initial guess $y_0(0)$ when solving for $y_0(h)$. The linear differential equation (5.2.13) is solved by using Euler's forward method (unmodified). Of the two alternatives for evaluating $\varepsilon x_1(0)$, we choose (5.2.22). These are very primitive numerical methods, but they are adequate to provide a comparison with conventional (non-stiff) methods.

The first comparison method is the modified Euler, the simplest Runge–Kutta method, applied to all equations of the system. The second comparison method is the trapezoidal rule which for the equation $y' = f(t, y)$, we take as

$$y^{(j)}(t + h) = y(t) + \frac{1}{2}h[f(t, y(t)) + f(t + h, y^{(j-1)}(t + h))],$$
$$j = 1, 2, \dots, 6.$$
$$y^{(0)}(t + h) = y(t).$$

This method may be viewed as the simplest prototype of stiff methods of

TABLE 5.2-1: two sets of calculations for the system of (5.2.24)

	t	Boundary layer method x	y	Exact solution x	y	Modified Euler x	y	Trapezoidal rule x	y
$h = 0.05$	0.0	1.0	1.0	1.0	1.0	1.0	1.0	1.0	1.0
$\theta = 0.05$	0.05	0.961	0.1	0.961	0.017	8.76E-1	8.43	0.937	-0.421
	0.10	0.915	0.01	0.915	0.01	3.65E-3	8.29E1	0.897	0.188
	0.15	0.871	0.01	0.871	0.01	-8.39	8.29 E2	0.851	-0.0733
	0.20	0.829	0.01	0.829	0.01	-9.21E1	8.30E3	0.81	0.0386
	0.25	0.789	0.01	0.789	0.01	-9.30E2	8.31E4	0.77	-0.00939
	0.30	0.751	0.01	0.751	0.01			0.732	0.0112
	0.35	0.715	0.01	0.715	0.01			0.696	0.00236
	0.40	0.681	0.01	0.680	0.01			0.662	0.00613
	0.45	0.648	0.01	0.648	0.01			0.629	0.00451
	0.50	0.617	0.01	0.617	0.01			0.598	0.00521
$h = 0.01$	0.0	1.0	1.0	1.0	1.0	1.0	1.0	1.0	1.0
$\theta = 0.1$	0.01	1.00	0.01	0.996	0.37	0.995	0.505	0.983	0.337
	0.02	0.990	0.01	0.989	0.14	0.990	0.503	0.971	0.116
	0.03	0.981	0.01	0.979	0.059	0.985	0.500	0.961	0.0419
	0.04	0.971	0.01	0.971	0.028	0.980	0.498	0.951	0.0173
	0.05	0.961	0.01	0.961	0.017	0.976	0.495	0.941	0.00909
	0.06	0.952	0.01	0.952	0.012	0.971	0.493	0.932	0.00636
	0.07	0.942	0.01	0.942	0.011	0.966	0.490	0.923	0.00545
	0.08	0.933	0.01	0.933	0.010	0.961	0.488	0.913	0.00515
	0.09	0.924	0.01	0.924	0.010	0.957	0.486	0.904	0.00505

Chapter 5

TABLE 5.2-2: two sets of calculations for the System of (5.2.25)

	t	Boundary layer method			Modified Euler			trapezoidal rule		
		x_1	x_2	x_3	x_1	x_2	x_3	x_1	x_2	x_3
$h = 0.05$	0.0	1.0	1.0	0.0	1.0	1.0	0.0	1.0	− 1.68	9.52
$\theta = 0.05$	0.05	1.01	1.02	1.18	1.00	1.01	− 1.31	1.00	1.02	− 1.71
	0.10	1.04	1.05	1.20	1.01	1.02	− 4.09	1.13	1.02	0.00
	0.15	1.06	1.07	1.22	1.02	1.04	− 10.0	2.08	1.13	− 8.92
	0.20	1.07	1.10	1.24	1.03	1.06	− 22.6	1.07	2.11	− 8.66
	0.25	1.09	1.12	1.27	1.07	1.12	− 49.5	9.42	1.08	− 7.56
	0.30	1.11	1.14	1.29	1.15	1.23	− 107.			
	0.35	1.13	1.17	1.31	1.33	1.45	− 228.			
	0.40	1.15	1.19	1.33	1.7	1.93	− 483.			
	0.45	1.17	1.22	1.35	2.51	2.97	− 1020.			
	0.50	1.19	1.24	1.37	4.32	5.27	− 2100.			
$\theta = 0.01$	0.0	1.0	1.0	0.0	1.0	1.0	0.0	1.00	1.00	0.00
$h = 0.01$	0.01	1.02	1.02	1.18	1.00	1.00	0.468	1.00	1.00	0.508
	0.02	1.04	1.04	1.20	1.00	1.00	0.747	1.00	1.00	0.793
	0.03	1.05	1.06	1.22	1.00	1.00	0.913	1.00	1.00	0.954
	0.04	1.07	1.08	1.24	1.00	1.00	1.01	1.00	1.00	1.04
	0.05	1.09	1.10	1.26	1.00	1.00	1.07	1.00	1.00	1.10
	0.06	1.11	1.11	1.28	1.00	1.01	1.11	1.00	1.00	1.12
	0.07	1.13	1.13	1.30	1.00	1.01	1.13	1.00	1.00	1.14
	0.08	1.14	1.15	1.32	1.00	1.01	1.14	1.00	1.01	1.15
	0.09	1.16	1.17	1.34	1.00	1.01	1.15	1.00	1.01	1.16

the A-stable or stiffly-stable type (see Chapter 2) so that it stands some-where in between a conventional method and a method devised for stiff equations.

Example 1. This example is the same as the one in Section 5.1.3, viz.

$$\dot{x} = y - x,$$
$$\dot{y} = 100y + 1. \qquad (5.2.24)$$

The numerical results are presented in Table 5.2-1.

Example 2. This example concerns the system

$$\dot{x}_1 = 0.0785(x_3 - x_1),$$
$$\dot{x}_2 = 0.1x_3, \qquad (5.2.25)$$
$$\dot{x}_3 = -(55 + x_2)x_3 + 65x_1.$$

The numerical results are presented in Table 5.2-2.

For additional computational results, see Aiken and Lapidus, 1974.

5.3. THE EXTRAPOLATION METHOD

We now show how the form of the perturbation solution can be used to calculate the stiff solution by combining solutions of auxiliary non-stiff or *relaxed equations*. This method which is called the *extrapolation method* begins by identifying a value of ε, say ε', which is substantially larger than ε in magnitude, but for which the solution of (5.1) with ε replaced by ε' can be used to approximate $x(h, \varepsilon)$, $y(h, \varepsilon)$. Thus, (5.1) is solved for larger values of ε, and so it can be solved more accurately with less effort. The number of operations used in these computations is proportional to $1/\varepsilon$ and $1/\varepsilon'$, respectively. Therefore the ratio ε'/ε provides a measure of the relative number of operations of direct solution (by some conventional numerical method) compared to the extrapolation method.

5.3.1. *Derivation of the Relaxed Equations*

We begin with an observation concerning the asymptotic form (5.1.1) and (5.1.2) of the solution to the initial value problem (5.1). From the boundary layer behavior of the inner solution (see (5.1.9) and (5.1.11)), we may specify two positive constants δ and K so that the solution of

(5.1) has the form

$$x(t, \varepsilon) = x_0(t) + x_1(t)\varepsilon + X(t/\varepsilon, \varepsilon) + O(\varepsilon^2),$$
$$y(t, \varepsilon) = y_0(t) + y_1(t)\varepsilon + Y(t/\varepsilon, \varepsilon) + O(\varepsilon^2). \tag{5.3.1}$$

Here X, Y have the following bounds.

$$\| X(t/\varepsilon, \varepsilon)\| + \| Y(t/\varepsilon, \varepsilon)\| \le Ke^{-\delta t/\varepsilon}. \tag{5.3.2}$$

δ is usually of the order of the smallest eigenvalue of g_y, and K depends on other data in the problem. These estimates hold uniformly for $0 \le t \le T$ and for all small positive ε. For details concerning this assertion see Hoppensteadt, 1971.

Now referring to (5.3.2), a value T is determined so that

$$Ke^{-\delta T} = O(h^{p+1}). \tag{5.3.3}$$

Clearly T is only determinable as an approximate value.

Next, a value $\varepsilon' = h/T$ is defined, and the system (5.1) is solved twice by a conventional (order p) integration method: first for $x(h, \varepsilon'/2), y(h, \varepsilon'/2)$, and then for $x(h, \varepsilon'), y(h, \varepsilon')$. It follows that

$$x(h, \varepsilon) = 2x(h, \varepsilon'/2) - x(h, \varepsilon') + O(h^{p+1}) + O((\varepsilon')^2) + O(\varepsilon),$$
$$y(h, \varepsilon) = 2y(h, \varepsilon'/2) - y(h, \varepsilon') + O(h^{p+1}) + O((\varepsilon')^2) + O(\varepsilon).$$

This relationship can be derived in the following way. From (5.1.1),

$$2x(h, \varepsilon'/2) = 2x_0(h) + x_1(h)\varepsilon' + 2X(2T, \varepsilon'/2) + O((\varepsilon')^2),$$
$$x(h, \varepsilon') = x_0(h) + x_1(h)\varepsilon' + X(T, \varepsilon') + O((\varepsilon')^2), \tag{5.3.4}$$

and so by subtracting,

$$2x(h, \varepsilon'/2) - x(h, \varepsilon') = x_0(h) + O(h^{p+1}) + O((\varepsilon')^2). \tag{5.3.5}$$

Here the term $O(h^{p+1})$ comes from estimating the boundary layer terms X in (5.3.2) by utilizing (5.3.3). On the other hand, by utilizing (5.3.1), (5.3.2) and (5.3.3), we find

$$x(h, \varepsilon) = x_0(h) + O(\varepsilon) + O((\varepsilon')^2),$$

and similarly, for $y(h, \varepsilon)$. Here $O(\varepsilon)$ is an estimate for the outer solution while $O((\varepsilon')^2)$ is, as in (5.3.5), an estimate for the boundary layer. The final result is that

$$x(h, \varepsilon) = 2x(h, \varepsilon'/2) - x(h, \varepsilon') + O(h^{p+1}) + O(\varepsilon) + O((h/T)^2),$$
$$y(h, \varepsilon) = 2y(h, \varepsilon'/2) - y(h, \varepsilon') + O(h^{p+1}) + O(\varepsilon) + O((h/T)^2). \tag{5.3.6}$$

We make the following two observations concerning this extrapolatory approach.

REMARK 5.3.1. The difference expressions $2x(h, \varepsilon'/2) - x(h, \varepsilon')$ and $2y(h, \varepsilon'/2) - y(h, \varepsilon')$ in (5.3.6) which are used as approximate replacements for $x(h, \varepsilon)$ and $y(h, \varepsilon)$, respectively, are in fact also approximate replacements for $x_0(h)$ and $y_0(h)$, respectively. Thus while calculating $x(h, \varepsilon')$ and $y(h, \varepsilon')$ from (5.1) is much easier than calculating $x(h, \varepsilon)$ and $y(h, \varepsilon)$, since $\varepsilon' \gg \varepsilon$, why do we not just calculate $x_0(h)$ and $y_0(h)$ from $(5.1.4_0)$? We give two answers.

(a) The exploitation of the methods of singular perturbation theory for the development of numerical techniques for stiff differential equations usually proceeds with the numerical determination of values of one or more terms in the asymptotic expansion supplied by that theory. The extrapolation formulas (5.3.6) break through this limitation of approach.

(b) Equation (5.1) with $\varepsilon = \varepsilon'$ is not stiff and may be easily and reliably solved by simple explicit numerical methods. While $(5.1.4_0)$ is also not stiff, employing it for the determination of $x_0(h)$ and $y_0(h)$ requires the solution of the nonlinear system $g = 0$ at each mesh point. This is usually a costly computation.

5.3.2. Computational Experiments

The following two computational experiments compare the extrapolation method (5.3.6) and the asymptotic expression itself.

(i) *A linear system*

We consider first the linear example studied analytically in Section 5.1.3 and numerically in Section 5.2.2:

$$\frac{dx}{dt} = y - x, \quad x(0) = \xi,$$

$$\frac{dy}{dt} = -\frac{y}{\varepsilon} + 1, \quad y(0) = \eta.$$

The eigenvalues of this system are -1 and $-1/\varepsilon \ll -1$. The exact solution is given by the formulas (see (5.1.14))

$$x(t) = e^{-t}\xi + (1 - e^{-t})\varepsilon - \left(\frac{\varepsilon}{1-\varepsilon}\right)(\eta - \varepsilon)(e^{-t/\varepsilon} - e^{-t}),$$

$$y(t) = \varepsilon + e^{-t/\varepsilon}(\eta - \varepsilon),$$

and the leading terms of the matched asymptotic expansion solutions are

$$x(t) = e^{-t}\xi + \varepsilon[(\eta - 1)e^{-t} + 1] + \cdots,$$
$$y(t) = \varepsilon + \cdots. \tag{5.3.7}$$

For any value of $\varepsilon \lesssim 10^{-5}$, the leading terms of the matched asymptotic expansion $(x_0(h), y_0(h))$ and the exact solution $(x(h, \varepsilon), y(h, \varepsilon))$ agree to about four figures. Thus while in Table 5.3–1, $\varepsilon = 10^{-5}$ is employed in the column labeled ε'/ε, the results in that table are otherwise valid for any $\varepsilon \lesssim 10^{-5}$. The results using the extrapolation method and evaluation of the matched asymptotic expansion (to leading order) are presented in Table 5.3-1. Since $K = 1 + |\eta|$ and $\delta = 1$ in this case, we take $T = -\ln[h^{p+1}/(1 + |\eta|)]$. In spite of the involved form of this formula for T, the values of the latter should be taken only approximately. The calculated values of $\varepsilon' = h/T$ are $\varepsilon' = 0.00819$ and $\varepsilon' = 0.00304$ corresponding to $h = 0.1$ and 0.01, respectively. Since these values of ε' are to be taken only as approximate, calculations for nearby values of ε' are also presented in the table.

TABLE 5.3-1:

$\xi = \eta = 1$ $p = 4$

		$h = 0.1$	$\varepsilon' = 0.00819$	
Extrapolation method	ε'	$x(h)$	$y(h)$	ε'/ε
	0.005	0.8632	0.0025	500
	0.00819	0.8648	0.004	819
	0.01	0.8657	0.005	1000
	0.02	0.8707	0.010	2000
Matched solution	——	0.9049	0.0	
		$h = 0.01$	$\varepsilon' = 0.00304$	
Extrapolation method	ε'	$x(h)$	$y(h)$	ε'/ε
	0.003	0.9866	0.00155	300
	0.00304	0.9866	0.00157	304
	0.004	0.9871	0.00255	400
Matched solution	——	0.9901	0.0	

Notice that the extrapolation method gives a 4% answer for $h = 0.1$, but it gives better than a 1% answer, the package for $h = 0.01$.

It is instructive to compare this extrapolation method with more typical stiff methods, such as the absolutely stable methods of Chapter 2. To do this, we invoke the latter by means of the frequently used package of C. W. Gear for integrating stiff differential equations. We use this package with $\varepsilon' = 10^{-5}$ for the second case in Table 5.3-1 with the following result. The package reaches $h = 0.01$ by employing a variable submesh of points which it determines adaptively. To produce a 1% answer, the package requires an average submesh size of 1.96×10^{-4}. The extrapolation method with $\varepsilon' = 0.00304$ produces its 1% answer with a fixed stepsize of 9.75×10^{-4}, nearly an order of magnitude difference. Of course as ε decreases the latter remains invariant, but the average stepsize employed by the package will decrease even further.

(ii) *A model enzyme reaction*

A simple enzyme reaction involves an enzyme E, substrate S, complex C and product P. Schematically, the reaction is

$$E + S \rightleftharpoons C, \qquad C \rightleftharpoons E + P.$$

After some preliminary scaling, this reaction can be described by a system of differential equations for the substrate concentration (x) and the complex concentration (y) as

$$\frac{dx}{dt} = -x + (x + k)y, \quad x(0) = 1,$$

$$\varepsilon \frac{dy}{dt} = x - (x + k')y, \quad y(0) = 0,$$

where ε measures a typical ratio of enzyme to substrate $(O(10^{-5}))$, and k and $k'(k < k')$ denote ratios of rate constants suitably normalized $(O(1))$.

The following Table 5.3-2 summarizes the result of these numerical calculations for $\varepsilon = 10^{-5}$ (although as noted in (i) above, except for the column ε'/ε, the results are valid for any $\varepsilon \lesssim 10^{-5}$), $h = 0.1$ and 0.1, $k = 1$, $k' = 2$. In this case, $K = 1$, $\delta = k'$, so we take $T = -((p + 1)/2)\ln h$. The calculated values for $\varepsilon' = h/T$ are $\varepsilon' = 0.04$ and $\varepsilon' = 0.0009$, respectively. Calculations are also presented for some nearby values of ε'.

The extrapolation method gives a 3% answer for $h = 0.1$, but it gives better than a 1% answer for $h = 0.01$. As in the case of Table 5.3-1, a

TABLE 5.3-2:

$x(0) = 1 \quad y(0) = 0 \quad p = 4$

		$h = 0.1$	$\varepsilon' = 0.04$	
Extrapolation method	ε'	$x(h)$	$y(h)$	ε'/ε
	0.01	0.9530	0.3229	1000
	0.04	0.9596	0.3247	4000
	0.05	0.9617	0.3253	5000
	0.1	0.9726	0.3285	10000
	0.15	0.9882	0.3350	15000
	0.2	0.9937	0.3406	20000
Matched solution	———	0.9888	0.3308	

		$h = 0.01$	$\varepsilon' = 0.0009$	
Extrapolation method	ε'	$x(h)$	$y(h)$	ε'/ε
	0.0004	0.9951	0.3322	40
	0.0008	0.9952	0.3323	80
	0.0009	0.9952	0.3323	90
	0.001	0.9952	0.3323	100
	0.0016	0.9954	0.3323	160
Matched solution	———	0.9917	0.3315	

comparison here produces an average step size of 2.4×10^{-4} for a 1% answer for Gear's package as opposed to a fixed stepsize of 2.7×10^{-3} for a 1% answer for the extrapolation method with $\varepsilon' = 0.0009$.

Note that a Runge–Kutta method of order $p = 4$ has been used to calculate the values of $x(h)$ and $y(h)$ furnished by the extrapolation method in all of the examples displayed in the Tables 5.3-1 and 5.3-2. In order to assure high accuracy and stability, these Runge–Kutta calculations were performed on a submesh of $[0, h]$ with a submesh increment $k = h\varepsilon'$.

Chapter 6

The Highly Oscillatory Problem

Summary

Except for the discussion involving aliasing in Section 4.3, we have up to now dealt with methods which exploit the transitory nature of the rapid change in solutions of stiff problems. Here we turn to the highly oscillatory problem wherein the rapid changes are expected to persist. A characteristic of the corresponding numerical methods to be discussed is the abandonment of the objective to obtain precise pointwise information about the solution.

We begin in Section 6.1 with a discussion of the two-time method, a technique for characterizing singular perturbation problems whose solutions are highly oscillatory. Numerical approximations to the solution which abandon precise phase information are then developed. The two-time technique requires computation of certain averages, and in Sections 6.2 and 6.3, we study this averaging procedure. Section 6.2 is concerned with an algebraic development of the averaging process, while in Section 6.3, we introduce an extrapolatory process for accelerating the computation of the average.

Finally, in Section 6.4, we adopt an approach to the highly oscillatory problem which consists of replacing the point functionals of standard numerical methods which indeed are unstable functionals for highly oscillatory stiff problems by other functionals which are stable. A numerical calculus is developed for the latter functionals in the context of the highly oscillatory problem.

6.1. A TWO-TIME METHOD FOR THE OSCILLATORY PROBLEM

In this section, we study the two-time technique for characterizing the asymptotic form of the solution of singular perturbation problems with highly oscillatory solutions. The asymptotic form is used to develop numerical approximations to the solution of the problem. The numerical approximations are characterized by the abandonment which is made of the need and desirability to obtain an approximation to the solution, in the traditional sense, i.e., pointwise. We begin this study with the introduction of the model problem for the case at hand.

6.1.1. *The Model Problem*

We consider the following model problem.

$$\varepsilon \frac{du}{dt} = (A + \varepsilon B)u, \qquad t \in (0, T],$$
$$u(0) = u_0,$$

(6.1.1)

where u is an n-vector and A and B are $n \times n$ matrices.

In terms of the *matrizant* $\Psi(t, \varepsilon)$, we write the solution of the model problem as

$$u = \Psi(t, \varepsilon)u_0,$$
$$\Psi(t, \varepsilon) = \exp\left[(A + \varepsilon B)t/\varepsilon\right].$$

(6.1.2)

If we introduce a new time scale

$$\tau = t/\varepsilon,$$

(6.1.3)

the solution becomes

$$u = e^{A\tau + Bt}u_0, \qquad 0 \leq \tau \leq T/\varepsilon.$$

(6.1.4)

This indicates that the solution develops on two different time scales: t called the *slow time* and τ called the *fast time*. If A and B commute, (6.1.4) becomes

$$u = e^{A\tau}e^{Bt}u_0.$$

(6.1.5)

In the commutative case the dependence on the two scales separates, and in principle, we could determine each of the factors in (6.1.5) separately and without computational difficulty.

However, in general, A and B do not commute, and moreover it is not necessarily the case that the development of the solution on the τ-scale is even meaningful to approximate numerically.

6.1.2. *Numerical Solution Concept*

Consider the example corresponding to

$$A = \begin{bmatrix} 0 & -1 \\ 1 & 0 \end{bmatrix}, \qquad B = \begin{bmatrix} -1 & 0 \\ 0 & -1 \end{bmatrix}.$$

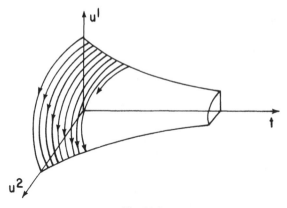

Fig. 6.1-1.

With these matrices, the motion described by (6.1.1) corresponds to a slowly damped (t-scale) extremely rapid (τ-scale) harmonic motion. The solution, indicated in Figure 6.1-1 for the case $n = 2$, is practically a surface filling curve.

As $\varepsilon \to 0$, the solution converges (in an approximate sense) to the cone obtained by rotating the curve $\| u_0 \| e^{-t}$ about the t-axis. Thus the meaningfulness of describing a trajectory by a set of its values on the points of a mesh is lost (i.e., is an ill-conditioned process).

A variety of alternate numerical solution concepts may be formulated. Consider the following one.

Numerical Solution Concept: Given $\varepsilon' > 0$ and $\delta > 0$, we say that $U(t)$ is an (ε', δ) *numerical approximation* to $u(t)$ at the time t, if there exists a positive $t' = t'(t)$ with $|t'| \leq \delta$ such that

$$\| U(t) - u(t + t') \| \leq \varepsilon'.$$

If t' is independent of t for t in some interval, then we say that $U(t)$ is a uniform (ε', δ) (numerical) approximation to $u(t)$.

Of course, $\delta = 0$ for the usual concept of (numerical) approximation. In Figure 6.1-2, an example of this approximation is given.

In terms of the model problem and by means of this approximation concept, we may accept a numerical approximation to the slow time part of the solution as a numerical approximation to the solution itself. The difficulty is to extract this part out of the whole solution, and to do this we

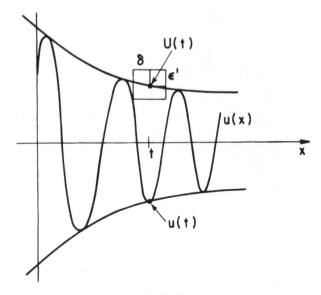

Fig. 6.1-2.

employ the *method of two-times*. (Of course nothing prevents us from computing the fast time part, as we will see, locally.) That is, to remove the ill-conditioning of the highly oscillatory problem, we must abandon some aspect of the solution, and in particular, we abandon the determination of its precise (fast time) phase. Pictorially speaking, imagine the solution to be a very tightly wound and slowly undulating helix. Imagine this helix to be cut at every mesh point and that each slice is shifted in turn by an unknown but very slight amount (i.e., piecewise discard the phase of the solution locally). What we seek is to determine numerical information about this mangled helix. In fact, unless a computation is performed on a scale of ε, which we certainly seek to avoid, we can only deal with this mangled helix. In terms of the approximation concept just introduced, the amount of information about the original helix which is contained in its mangled version is acceptable and adequate when ε is small.

6.1.3. *The Two-time Expansion*

We seek approximations to the solution of (6.1.1) in the form of a general *two-time expansion:*

$$u = \sum_{r=0}^{\infty} u_r(t, \tau)\varepsilon^r. \tag{6.1.6}$$

This will be a useful series for purposes of approximation, if we have

$$\boldsymbol{u}_r(t, t/\varepsilon)\varepsilon^r = o(\varepsilon^{r-1}), \qquad r = 1, 2, \ldots, \tag{6.1.7}$$

as $\varepsilon \to 0$, uniformly for $0 \le t \le T$. With (6.1.7) valid, we say that (6.1.6) is an *asymptotic expansion with asymptotic scale* ε. A sufficient condition for (6.1.7) is that

$$\boldsymbol{u}_r(t, \tau) = o(\tau), \tag{6.1.8}$$

as $\tau \to \infty$, for $r = 1, 2, \ldots$.

The expansion resulting from this prescription of the form (6.1.6)–(6.1.8) of the solution will be derived below. It is sometimes possible to obtain more information from the expansion by placing a stronger condition than (6.1.8) on the coefficients. In particular, we will determine conditions on A and B so that the requirement

$$\boldsymbol{u}_r(t, \tau) = o(\tau e^{A\tau}), \tag{6.1.9}$$

as $\tau \to \infty$ for $r = 1, 2, \ldots$, can be used to obtain a valid expansion.

If A is an *oscillatory matrix* (all its eigenvalues have zero real parts), the conditions (6.1.8) and (6.1.9) are equivalent. If A is a *stable matrix* (all its eigenvalues have negative real parts), the condition (6.1.9) is more restrictive than (6.1.8). In the stable case, it may not be possible to obtain an expansion of the solution of (6.1.1) in the form (6.1.6) wherein the coefficients satisfy either (6.1.8) or (6.1.9). However, we will describe another restriction on the problem which when used with (6.1.9) guarantees that the solution of (6.1.1) can be approximately solved in the form (6.1.6). This approximation technique proceeds via the two-time approach. This result is valid when the eigenvalues of A lie in the stable half plane; therefore, it contains both the stable and oscillatory cases. In the stable case, the expansion found by this method reduces to the one which would be obtained by the method of matched asymptotic expansions (see Section 5.1). In the oscillatory case, this result reduces to an expansion equivalent to the one obtained by the Bogoliubov *method of averaging* (see Volosov, 1962).

6.1.4. *Formal Expansion Procedure*

We consider the initial value problem for the system (6.1.1), and we write the initial conditions in the form

$$\boldsymbol{u}(0) = \sum_{r=0}^{\infty} \boldsymbol{a}_r \varepsilon^r. \tag{6.1.10}$$

To simplify computation, let

$$v(t, \tau) = e^{-A\tau} u(t, \tau).$$ (6.1.11)

Since v is considered as a function of the two variables τ and $t = \varepsilon\tau$,

$$\frac{dv(\varepsilon\tau, \tau)}{d\tau} = \varepsilon \frac{\partial v(t, \tau)}{\partial t} + \frac{\partial v(t, \tau)}{\partial \tau}.$$ (6.1.12)

Then (6.1.1) becomes the following equation for v.

$$\varepsilon \frac{\partial v}{\partial t} + \frac{\partial v}{\partial \tau} = \varepsilon B(\tau)v, \qquad v(0) = \sum_{r=0}^{\infty} a_r \varepsilon^r,$$ (6.1.13)

where

$$B(\tau) = e^{-A\tau} Be^{A\tau}.$$ (6.1.14)

We seek a solution in the form (6.1.6) which becomes

$$v = \sum_{r=0}^{\infty} v_r(t, \tau)\varepsilon^r,$$ (6.1.15)

subject to the condition (6.1.9) on the u_r. In terms of the v_r, this condition becomes

$$v_r(t, \tau) = o(\tau) \quad \text{as} \quad \tau \to \infty, \qquad r = 0, 1, \ldots.$$ (6.1.16)

Substituting (6.1.15) into (6.1.13) and equating coefficients of the like powers of ε gives

$$\frac{\partial v_r}{\partial \tau} = B(\tau)v_{r-1} - \frac{\partial v_{r-1}}{\partial t}, \quad v_r(0,0) = a_r, \quad r = 0, 1, \ldots,$$ (6.1.17)

Here $v_{-1} \equiv 0$.

The problem (6.1.17) is underdetermined. The equation (6.1.17) for v_r can be integrated to give

$$v_r(t, \tau) = \tilde{v}_r(t) + \int_0^\tau \left[B(\sigma)v_{r-1}(t, \sigma) - \frac{\partial v_{r-1}(t, \sigma)}{\partial t} \right] d\sigma, \qquad r = 0, 1, \ldots,$$ (6.1.18)

where

$$\tilde{v}_r(0) = a_r.$$ (6.1.19)

Except for (6.1.19), $\tilde{v}_r(t)$ is arbitrary. Differentiating (6.1.18) with respect to

t gives

$$\frac{\partial v_r}{\partial t} = \frac{\partial \tilde{v}_r}{\partial t} + \int_0^\tau \left[B(\sigma) \frac{\partial v_{r-1}}{\partial t} - \frac{\partial^2 v_{r-1}}{\partial t^2} \right] d\sigma. \tag{6.1.20}$$

Combining this with (6.1.18) gives

$$v_r(t, \tau) = \tilde{v}_r(t) + \int_0^\tau B(\sigma) d\sigma \tilde{v}_{r-1}(t) - \tau \frac{d\tilde{v}_{r-1}}{dt} + \int_0^\tau R_{r-1}(t, \sigma) d\sigma, \tag{6.1.21}$$

where

$$R_r(t, \sigma) = -\int_0^\sigma \left[B(\sigma) \frac{\partial v_{r-1}(t, \sigma')}{\partial t} - \frac{\partial^2 v_{r-1}(t, \sigma')}{\partial t^2} \right] d\sigma'$$

$$+ B(\sigma) \int_0^\sigma \left[B(\sigma') v_{r-1}(t, \sigma') - \frac{\partial v_{r-1}(t, \sigma')}{\partial t} \right] d\sigma'. \tag{6.1.22}$$

(6.1.21) and (6.1.22) hold for $r = 0, 1, \ldots$, where $\tilde{v}_{-1} \equiv R_0 \equiv 0$. Let us impose the growth condition (6.1.16) in (6.1.21). To do this divide (6.1.21) by τ and take the limit as $\tau \to \infty$. This results in the following condition for $\tilde{v}_{r-1}$.

$$\frac{d\tilde{v}_{r-1}}{dt} = \left(\lim_{\tau \to \infty} \frac{1}{\tau} \int_0^\tau B(\sigma) d\sigma \right) \tilde{v}_{r-1}$$

$$+ \lim_{\tau \to \infty} \frac{1}{\tau} \int_0^\tau R_{r-1}(t, \sigma) d\sigma, \quad r = 0, 1, \ldots. \tag{6.1.23}$$

When these limits exist, (6.1.23) along with (6.1.19) determine $\tilde{v}_r, r = 0, 1, \ldots$.

This approach depends critically on the existence of the limits in (6.1.23). The development will be simplified by using the following notation.

$$\bar{f} = \lim_{\tau \to \infty} \frac{1}{\tau} \int_0^\tau f(\sigma) d\sigma. \tag{6.1.24}$$

If $\bar{f}$ exists we will call it the *average of f*. In terms of this notation, (6.1.23) becomes

$$\frac{d\tilde{v}_r}{dt} = \bar{B} \tilde{v}_r + \bar{R}_r(t), \quad \tilde{v}_r(0) = a_r, \quad r = 0, 1, \ldots, \tag{6.1.25}$$

provided the averages exist.

6.1.5. *Existence of the Averages and Estimates of the Remainder*

In this section, we study the existence of the averages appearing in (6.1.25). We content ourselves with a study of $\bar{B}$ and $\bar{R}_1$ since it is these two averages which provide for the existence of the approximation $v_0 + \varepsilon v_1$ to v. This approximation is adequate for our computational purposes. We conclude this section with an estimate of the quality of the approximation of v by $v_0 + \varepsilon v_1$, an implicit estimate of the remainder $\boldsymbol{R}_1$.

Existence of the Average, $\bar{B}$

The example $A = \begin{bmatrix} -1 & 0 \\ 0 & -2 \end{bmatrix}$, $B = \begin{bmatrix} 0 & 0 \\ 1 & 0 \end{bmatrix}$ shows that $\bar{B}$ may not exist.

We assume in this analysis that the matrix A has simple elementary divisors. In this case, we may assume without loss of generality that A is a diagonal matrix. We denote the elements of A and B, respectively, by

$$A = (\lambda_i \delta_{ij}) \quad \text{and} \quad B = (b_{ij}). \tag{6.1.26}$$

Then

$$\frac{1}{\tau} \int_0^\tau e^{-A\sigma} B e^{A\sigma} = (b_{ij} f_{ij}), \tag{6.1.27}$$

where

$$f_{ij} = \begin{cases} \dfrac{e^{(\lambda_j - \lambda_i)\tau} - 1}{\tau(\lambda_j - \lambda_i)} & , \quad \lambda_i \neq \lambda_j, \\ 1, & \quad \lambda_i = \lambda_j. \end{cases} \tag{6.1.28}$$

This computation demonstrates the following theorem.

THEOREM 6.1.1. $\bar{B}$ *exists if and only if the elements b_{ij} of B vanish whenever* Re $(\lambda_j - \lambda_i) > 0$.

We henceforth assume that the hypothesis of Theorem 6.1.1 holds. The computation also has the following corollary.

Corollary 6.1.2
   (i)   *If $\bar{B}$ exists, then $\bar{B}_{ij} = b_{ij} \delta(\lambda_i, \lambda_j)$, where δ is the Kronecker delta.*
   (ii) *If the eigenvalues of A are distinct, then $\bar{B} = (b_{ij} \delta_{ij})$.*
   (iii) *If A and B are normal matrices, then $\bar{B}$ exists and is conjugate to B.*
   (iv) *If A and B commute, then $\bar{B}$ exists and is conjugate to B.*

(v) *If B is diagonal, then $\bar{B} = B$.*

(vi) *If $A = \lambda I$, then $\bar{B} = B$.*

(vii) *Let $\bar{B}$ exist and suppose that $\lambda_1 = \lambda_2 = \ldots = \lambda_m = 0$ and $\lambda_{m+i} \neq 0$, $i = 1, \ldots, n - m$. Let B_{11} be the $m \times m$ principle submatrix of B. Then the $m \times m$ principle submatrix of $\bar{B}$ is B_{11} while $\bar{B}_{ij} = 0$ for $j > m$ and $i \le m$ and for $j \le m$ and $i > m$.*

In terms of a block decomposition of B and $\bar{B}$, (vii) may be represented by:

$$B = \begin{bmatrix} B_{11} & B_{12} \\ B_{21} & B_{22} \end{bmatrix} \Rightarrow \bar{B} = \begin{bmatrix} B_{11} & 0 \\ 0 & \bar{B}_{22} \end{bmatrix},$$

where $\bar{B}_{22}$ is unspecified but assumed to exist.

Proof. (iii) and (iv) follow since A and B are simultaneously diagonalizable in these cases. All other statements are immediate. $\qquad\square$

Existence of the average $\bar{R}_1$

The following theorem characterizes the existence of $\bar{R}_1$.

THEOREM 6.1.3. *$\bar{R}_1$ exists whenever $\bar{B}$ exists.*

Proof. The proof follows from Theorem 6.1.1 and the following computation. Let

$$\rho(\sigma) = \int_0^\sigma (B(\sigma') - \bar{B})d\sigma'. \tag{6.1.29}$$

Then from (6.1.22), we have

$$R_1(t, \sigma) = (B(\sigma)\rho(\sigma) - \rho(\sigma)\bar{B})e^{\bar{B}t}v_0(0), \tag{6.1.30}$$

where we have used $d\tilde{v}_0/dt = \bar{B}\tilde{v}_0$, the latter following from (6.1.25).

We first show the existence of $\bar{\rho}$. From (6.1.28), we see that the ijth element of ρ is

$$\rho_{ij} = \left[b_{ij} \times \begin{cases} \dfrac{e^{(\lambda_j - \lambda_i)\sigma} - 1}{\lambda_j - \lambda_i}, & \lambda_i \neq \lambda_j \\ \sigma, & \lambda_i = \lambda_j \end{cases} \right] - \begin{bmatrix} 0, & \lambda_i \neq \lambda_j \\ \sigma b_{ij}, & \lambda_i = \lambda_j \end{bmatrix}$$

$$= \begin{bmatrix} b_{ij} \dfrac{e^{(\lambda_j - \lambda_i)\sigma} - 1}{\lambda_j - \lambda_i}, & \lambda_i \neq \lambda_j \\ 0, & \lambda_i = \lambda_j \end{bmatrix}.$$

Then

$$
\frac{1}{\tau}\int_0^\tau \rho(\sigma)d\sigma =
\begin{bmatrix}
\dfrac{b_{ij}}{\lambda_j - \lambda_i}\left[\dfrac{1}{\tau}\displaystyle\int_0^\tau e^{(\lambda_j - \lambda_i)\sigma}d\sigma - 1\right], & \lambda_i \neq \lambda_j \\[4mm]
0, & \lambda_i = \lambda_j
\end{bmatrix}. \quad (6.1.31)
$$

The limit of (6.1.31) as $\tau \to \infty$ exists if and only if $b_{ij} = 0$ whenever $\mathrm{Re}(\lambda_j - \lambda_i) > 0$. In this case we have

$$
\bar{\rho} =
\begin{bmatrix}
\dfrac{-b_{ij}}{\lambda_j - \lambda_i}, & \lambda_i \neq \lambda_j \\[4mm]
0, & \lambda_i = \lambda_j
\end{bmatrix}, \quad (6.1.32)
$$

demonstrating the existence of $\bar{\rho}$.

We now show the existence of the average $\overline{B\rho}$. We have

$$
\begin{aligned}
(B\rho)_{ij} &= \sum_{k=1}^n B_{ik}\rho_{kj} = \sum_{k=1}^n b_{ik}e^{(\lambda_k - \lambda_i)\sigma}(\sigma b_{kj}f_{kj} - \sigma\bar{B}_{kj}) \\
&= \sum_{\substack{k=1 \\ \lambda_j \neq \lambda_k}}^n b_{ik}b_{kj}\frac{e^{(\lambda_j - \lambda_i)\sigma} - e^{(\lambda_k - \lambda_i)\sigma}}{\lambda_j - \lambda_k} + \sum_{\substack{k=1 \\ \lambda_j = \lambda_k}}^n (\sigma b_{ik}b_{kj} - \sigma b_{ik}\bar{B}_{kj}) \\
&= \sum_{\substack{k=1 \\ \lambda_j \neq \lambda_k}}^n b_{ik}b_{kj}\frac{e^{(\lambda_j - \lambda_i)\sigma} - e^{(\lambda_k - \lambda_i)\sigma}}{\lambda_j - \lambda_k} \\
&= \sum_{\substack{k=1 \\ \lambda_j \neq \lambda_k}}^n b_{ik}e^{(\lambda_k - \lambda_i)\sigma}b_{kj}\frac{e^{(\lambda_j - \lambda_i)\sigma} - 1}{\lambda_j - \lambda_k}.
\end{aligned}
$$

If the real part of any exponent appearing in this expression is positive, then by hypothesis the corresponding element of B appearing in front of that exponential term vanishes. Thus the sum appearing here contains only exponentials with exponents with non-positive real part. Thus $\overline{B\rho}$ exists.

This completes the proof of the theorem.   □

Using the techniques of this proof, we may deduce the following relation, which will be exploited in the estimate of the remainder to follow.

$$
(\overline{B\rho} - \bar{\rho}\bar{B})_{ij} = \sum_{\substack{k=1 \\ \lambda_k \neq \lambda_i}}^n \frac{b_{ik}\bar{B}_{kj}}{\lambda_k - \lambda_i}. \quad (6.1.34)
$$

Estimate of the remainder
Let

$$w = v - v_0 - \varepsilon v_1. \tag{6.1.35}$$

As we remarked above, we restrict ourselves to estimate w. We will derive a differential equation for w, solve this equation and estimate its solution.

Using $dv/d\tau = \varepsilon B(\tau)v$ (see (6.1.13)) and $dv_0/dt = \bar{B}v_0$ (see (6.1.21)) and (6.1.25)), we differentiate (6.1.35) to obtain

$$
\begin{aligned}
dw(\varepsilon\tau, \tau)/d\tau &= \varepsilon B v(\varepsilon\tau, \tau) - \varepsilon\bar{B}v_0(\varepsilon\tau) - \varepsilon dv_1(\varepsilon\tau, \tau)/d\tau \\
&= \varepsilon B(w + v_0 + \varepsilon v_1) - \varepsilon\bar{B}v_0(\varepsilon\tau) - \varepsilon dv_1(\varepsilon\tau, \tau)/d\tau.
\end{aligned}
\tag{6.1.36}
$$

Let

$$C(\sigma) = B(\sigma)\rho(\sigma) - \rho(\sigma)\bar{B}. \tag{6.1.37}$$

Then from (6.1.25) and (6.1.30), we obtain

$$d\tilde{v}_1(t)/dt = \bar{B}\tilde{v}_1 + \bar{C}v_0(t). \tag{6.1.38}$$

From (6.1.18) and (6.1.29), we have

$$v_1(t, \tau) = \tilde{v}_1(t) + \rho(\tau)v_0(t). \tag{6.1.39}$$

Differentiating this relation with respect to τ gives

$$\frac{dv_1(\varepsilon\tau, \tau)}{d\tau} = \varepsilon\frac{d\tilde{v}_1(\varepsilon\tau)}{dt} + (B(\tau) - \bar{B})v_0(\varepsilon\tau) + \varepsilon\rho(\tau)\bar{B}v_0(\varepsilon\tau). \tag{6.1.40}$$

Combining this with (6.1.38) gives us an expression for $dv_1(\varepsilon\tau, \tau)/d\tau$ which when inserted into (6.1.36) gives us the following differential equation for w.

$$
\begin{aligned}
\frac{d}{d\tau}w(\varepsilon\tau, \tau) &= \varepsilon B(w + v_0 + \varepsilon v_1) - \varepsilon\bar{B}v_0 - \varepsilon^2\bar{B}\tilde{v}_1 - \varepsilon^2\bar{C}v_0 \\
&\quad - \varepsilon(B - \bar{B})v_0 + \varepsilon^2\rho\bar{B}v_0.
\end{aligned}
\tag{6.1.41}
$$

Note that $\bar{C}$ is given by (6.1.34). Using (6.1.37) and (6.1.39), (6.1.41) becomes

$$\frac{d}{d\tau}w(\varepsilon\tau, \tau) = \varepsilon Bw + \varepsilon^2[(B - \bar{B})\tilde{v}_1 + (C - \bar{C})\tilde{v}_0]. \tag{6.1.42}$$

To estimate w, we first introduce the fundamental matrix Ψ defined by

$$\frac{d\Psi}{d\tau} = \varepsilon B\Psi, \quad \Psi(0) = I. \tag{6.1.43}$$

Note that

$$\frac{d}{d\tau}(\Psi^{-1}) = -\varepsilon\Psi^{-1}B. \tag{6.1.44}$$

Let $B*$ denote the complex conjugate of B and B^T its transpose. Then

$$\frac{d}{d\tau}\|\Psi\|^2 \equiv \frac{d}{d\tau}(\Psi^T\Psi*) = \varepsilon\Psi^T(B^T + B*)\Psi*. \tag{6.1.45}$$

Then since $B^T + B*$ is Hermitian, there exists a constant K (e.g., the magnitude of the largest eigenvalue of $B^T + B*$) such that

$$\frac{d}{d\tau}\|\Psi\|^2 \leq \varepsilon K\|\Psi\|^2. \tag{6.1.46}$$

Then

$$\|\Psi\|^2 \leq e^{\varepsilon K\tau}. \tag{6.1.47}$$

By our hypothesis (see Theorem 6.1.1f.), $B(\tau)$ is a bounded function of τ for $\tau \geq 0$. Thus K is independent of τ, and (6.1.47) shows that $\|\Psi\|^2$ is bounded unformily for τ restricted to an interval of the form $[0, T/\varepsilon]$.

Now we solve (6.1.42), and write

$$w(\varepsilon\tau, \tau) = \Psi(\tau)[a_2\varepsilon^2 + \ldots]$$
$$+ \varepsilon^2 \int_0^\tau \Psi(\tau)\Psi^{-1}(\sigma)[(B - \bar{B})\tilde{v}_1(\varepsilon\sigma) + (C - \bar{C})\tilde{v}_0(\varepsilon\sigma)]d\sigma. \tag{6.1.48}$$

In (6.1.48) write $(B - \bar{B})$ as $(d/d\sigma)\int_0^\sigma (B - \bar{B})d\sigma'$ and $(C - \bar{C})$ as $(d/d\sigma)$ $\int_0^\sigma (C - \bar{C})d\sigma'$. Then use the relation $(FGH)' = FGH' + FG'H + F'GH$ to integrate by parts the integral with respect to σ'. Using (6.1.29) and (6.1.44) in what results, we finally obtain the following representation of w.

$$w(\varepsilon\tau, \tau) = \varepsilon^2\{\Psi(\tau)O(1) + \rho(\tau)\tilde{v}_1(\varepsilon\tau) + \int_0^\tau (C - \bar{C})d\sigma'\tilde{v}_0(\varepsilon\tau)$$
$$+ \varepsilon \int_0^\tau \Psi(\tau)\Psi^{-1}(\sigma)[B(\sigma)\int_0^\sigma (C - \bar{C})d\sigma'\tilde{v}_0(\varepsilon\sigma)$$
$$- \rho(\sigma)\frac{d}{dt}\tilde{v}_1(\varepsilon\sigma) + B\rho(\sigma)\tilde{v}_1(\varepsilon\sigma)$$
$$- \int_0^\sigma (C - \bar{C})d\sigma'\bar{B}\tilde{v}_0(\varepsilon\sigma)]d\sigma\}. \tag{6.1.49}$$

We have noted that $\| \Psi \|$ is bounded uniformly for τ in the interval $[0, T/\varepsilon]$ and that B is bounded for $\tau \geq 0$ (see (6.1.47)f.). The functions $\tilde{v}_0(\varepsilon\tau)$ and $\tilde{v}_1(\varepsilon\tau)$ appearing in (6.1.49) are defined as continuous functions and their arguments range over the bounded interval $0 \leq \varepsilon\tau \leq T$. Thus these functions are uniformly bounded. The quantities $\rho(\sigma)$ and $\int_0^\sigma (C - \bar{C})d\sigma'$ appearing in (6.1.49) exist as bounded functions for $0 \leq \sigma \leq \infty$ since they are continuous functions. This accounts for every quantity appearing in (6.1.49). Thus we may conclude that $\| w \|$ is bounded, and the estimate of the remainder is complete. We summarize this estimate with the statement of the following theorem.

THEOREM 6.1.4. $\displaystyle\max_{0 \leq \tau \leq T/\varepsilon} \left| v(\varepsilon\tau, \tau) - v_0(\varepsilon\tau) - \varepsilon v_1(\varepsilon\tau, \tau) \right| \leq \text{const } \varepsilon^2.$

For a more complete treatment of (6.1.1) by the two-time method in the general case where the eigenvalues of A may be anywhere in the complex plane and where nonlinear forcing terms are adjoined to the system as well, see Hoppensteadt and Miranker, 1976.

6.1.6. The Numerical Algorithm

We take the leading term, $u_0(t, \tau)$ of the expansion (6.1.6) as an approximation to the solution of the initial value problem (6.1.1) with the initial condition given in (6.1.10).

Then from (6.1.11) and (6.1.18)

$$u_0(t, \tau) = \Phi(\tau)\tilde{v}_0(t). \tag{6.1.50}$$

$\Phi(\tau)$ is the *fundamental matrix* given by

$$\Phi_\tau = A\Phi, \qquad \Phi(0) = I, \tag{6.1.51}$$

where I is the $n \times n$ identity matrix, while from (6.1.25)

$$d\tilde{v}_0/dt = \bar{B}\tilde{v}_0, \qquad \tilde{v}_0(0) = a_0. \tag{6.1.52}$$

From (6.1.14)

$$\bar{B} = \lim_{\tau \to \infty} \frac{1}{\tau} \int_0^\tau \Phi^{-1}(\sigma)B\Phi(\sigma)d\sigma. \tag{6.1.53}$$

We describe the algorithm for replacing a_0, the approximation to $u(0)$, by $U(h)$, the approximation to $u(h)$, (in the sense of the approximation concept in Section 6.1.2 above). The algorithm is to be repeated approximating $u(t)$ at $2h, 3h, \ldots$, successively.

Algorithm

(i) Solve (6.1.51) on a mesh of increment k in the τ-scale by some self starting numerical method, obtaining the sequence $\Phi(jk)$, $j = 0, \ldots, N$.

(ii) Using the values $\Phi(jk)$ obtained in (i), approximating $\bar{B}$ by truncating the limit of τ integration and replacing the integral in (6.1.53) by a quadrature formula, say

$$\bar{B} = \frac{1}{N} \sum_{j=0}^{N} c_k \Phi^{-1}(jk)B\Phi(jk).$$

The integer N is determined by a numerical criterion which insures that the elements of the matrix $\bar{B}$ are calculated to some desired accuracy.

(iii) With $\bar{B}$ (approximately) determined in (ii), solve (6.1.52) for $\tilde{v}_0(h)$ by some self-starting numerical method.

(iv) Compute $u_0(h, Nk) = \Phi(Nk)\tilde{u}_0(h)$, and take this as the approximation to $u(h)$.

Refinement

The method may be refined by adding an approximation of $\varepsilon v_1(h, h/\varepsilon)$ to $\tilde{v}_0(h)$ prior to multiplication by $\Phi(Nk)$ (step (iv)). This approximation in turn is determined from a numerical solution of the equations defining $v_1(t, \tau)$, viz.

$$v_1(t, \tau) = \tilde{v}_1(t) - \left(\bar{B}t + \int_0^\tau B(\sigma)d\sigma \right)e^{\bar{B}t}a_0,$$

$$d\tilde{v}_1/dt = \bar{B}\tilde{v}_1 + \bar{R}_1(t), \quad \tilde{v}_1(0) = a_1,$$

$$R_1(t, \sigma) = \left[(\bar{B}^2 - B(\sigma)\bar{B})\sigma - \int_0^\sigma B(\sigma')d\sigma'\bar{B} \right.$$
$$\left. + B(\sigma)\int_0^\sigma B(\sigma')d\sigma' \right]\tilde{v}_0(t).$$

In Figure 6.1-3, we schematize the computation. Of course, in practice ε will be extremely small so that unlike the schematic an enormous number of oscillations of Φ will occur in the t interval $[0, h]$. Notice how far the computed answer $\Phi(Nk)\tilde{v}_0(h)$ may be from the usual approximation to the solution, $u_0(h, h/\varepsilon)$.

The fundamental matrix $\Phi(\tau)$ is composed of modes corresponding to the eigenvalues of A. Since the eigenvalues of A lie in the closed left half plane, the profile for (a component of) Φ will, after some moderate

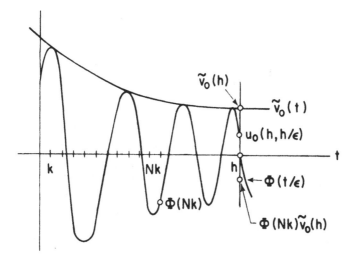

Fig. 6.1-3.

number of cycles, settle down to an (almost-)periodic function. Thus the set of mesh points $\{jk \mid j = 0, \ldots, N\}$ may be expected to extend over just these cycles (approximately).

6.1.7. Computational Experiments

In Table 6.1-1, we give the results of calculations with three sample problems, $P_i, i = 1, 2, 3$. P_1 corresponds to a damped case (A has real eigenvalues), P_2 to a purely oscillatory A and P_3 to a mixed case. The

TABLE 6.1-1

Problem P_1 (damped case)

$$A = \begin{bmatrix} 0 & 0 \\ 0 & -1 \end{bmatrix} \quad ' B = \begin{bmatrix} -1 & 1 \\ 0 & 0 \end{bmatrix} \quad \varepsilon = 0.01, h = 0.05, k = 0.05$$

t	$\tilde{v}_0$		$\Phi(kN)\tilde{v}_0$	
0.0	1.000	1.00	1.000	1.00
0.05	0.953	1.00	0.953	0.0
0.10	0.908	1.00	0.906	0.0
0.15	0.865	1.00	0.862	0.0
0.20	0.824	1.00	0.820	0.0
0.25	0.785	1.00	0.780	0.0

TABLE 6.1-1 (*Contd.*)

Problem P₂ (oscillatory case)

$$A = \begin{bmatrix} 0 & -1 \\ 1 & 0 \end{bmatrix} \quad B = \begin{bmatrix} -2 & 0 \\ 0 & 0 \end{bmatrix} \quad \varepsilon = 0.001, h = 0.01, k = 0.05$$

t	$\tilde{v}_0$		$\Phi(kN)\tilde{v}_0$	
0.0	0.500	0.500	0.500	0.500
0.01	0.495	0.495	0.325	0.605
0.02	0.490	0.490	0.184	0.669
0.03	0.485	0.485	0.007	0.687
0.04	0.480	0.481	−.167	0.660
0.05	0.475	0.476	−.327	0.589

Problem P₃ (mixed case)

$$A = \begin{bmatrix} 0 & 0 & 0 & 0 \\ 0 & -1 & 0 & 0 \\ 0 & 0 & 1 & -1 \\ 0 & 0 & 0 & 0 \end{bmatrix} \quad B = \begin{bmatrix} 1 & 1 & 1 & 0 \\ 0 & 1 & 0 & 0 \\ 1 & 1 & 1 & 0 \\ 0 & 0 & 0 & 1 \end{bmatrix} \quad \varepsilon = 0.01, h = 0.05, k = 0.05$$

t	$\tilde{v}_0$				$\Phi(kN)\tilde{v}_0$			
0.0	1.00	1.00	1.00	1.00	1.00	1.0	1.000	1.000
0.05	1.05	1.05	1.06	1.04	1.05	0.0	− 1.450	0.327
0.10	1.11	1.11	1.12	1.09	1.10	0.0	0.534	− 1.460
0.15	1.17	1.16	1.18	1.14	1.17	0.0	0.997	1.310
0.20	1.23	1.22	1.24	1.19	1.22	0.0	− 1.720	0.149
0.25	1.29	1.28	1.31	1.24	1.28	0.0	0.846	− 1.600

numerical methods used are chosen to be the most elementary (e.g., Euler's method for differential equations and Simpson's rule for integrals) so that the results are accurate only to a few percent. Moreover $\varepsilon/h = 0.1$ or 0.2 so that the examples are not particularly stiff. Of course this means that the examples are not ones which would show the methods at their best, since the stiffer the problem the better are these methods.

6.2. ALGEBRAIC METHODS FOR THE AVERAGING PROCESS

We study the algebraic aspects of the averaging process (6.1.53). This will lead to a solution of the problem of computing the average (6.1.53) in the sense that we will provide several alternative ways for giving $\bar{B}$.

6.2.1. Algebraic Characterization of Averaging

We begin this algebraic study by showing that the integrand, $e^{-A\sigma}Be^{A\sigma}$ which is being averaged in (6.1.53) spans a subspace S, of n^2-dimensional Euclidean space. To do this we use three algebraic constructs or mappings called exp, ad and Ad, respectively. We describe a base for and then invariant properties of S. This leads to Theorem 6.2.3 which gives the value of the average $\bar{B}$.

The mappings exp, ad *and* Ad

Let $E(n)$ denote the set of $n \times n$ matrices considered as a vector space over the field of definition which will be either R, the reals: or C, the complex numbers. Let $GL(n) \subseteq E(n)$ be the *group of nonsingular linear transformations*.

Let GL denote the *group of nonsingular linear transformations* of $E(n)$, and let E denote the *space of linear endormorphisms of $E(n)$*. Clearly E can be identified with the space of $n^2 \times n^2$ matrices.

For $A \in E(n)$ (or for $A \in E$), we define $\exp(A) \equiv e^A$ as

$$\exp(A) = I + A + \frac{A^2}{2!} + \dots. \tag{6.2.1}$$

Clearly $\exp(A) \in GL(n)$ or $\exp(A) \in GL$ as the case may be.

For future reference, we note that if a linear vector space V has the property that $A(V) \subseteq V$, then $\exp(tA)(V) \subseteq V$, $\forall t \in R$.

It is well-known that $\exp(tA)$, $t \in R$, is a one parameter subgroup of $GL(n)$ (or of GL) and that

$$\frac{d}{dt}\exp(tA)\big|_{t=0} = A.$$

Further, for any one parameter subgroup $g(t)$ of $GL(n)$, we have that if

$$\frac{d}{dt}g(t)\big|_{t=0} = A, \quad \text{then} \quad g(t) = \exp(tA).$$

The form of the integrand in (6.1.53) leads us to the concept of the *adjoint representations.* For $g \in GL(n)$, we define ad(g) by the formula

$$\text{ad}(g)(X) = gXg^{-1}, \quad \forall x \in E(n).$$

Multiplication in the right member here denotes matrix multiplication. That ad(g)$\in E$ comes from the following relation which expresses a standard property of matrix multiplication.

$$g(a_1 X_1 + a_2 X_2)g^{-1} = a_1 g X_1 g^{-1} + a_1 g X_2 g^{-1}.$$

Here a_1, $a_2 \in R$ or C and $X_1, X_2 \in E(n)$. Note also that since ad($g_1 g_2$) = ad(g_1) ad(g_2), then ad(g^{-1}) is the inverse of ad(g). Thus ad(g)$\in GL$.

We will make use of an additional mapping denoted by Ad(A) and defined as follows. Let $A \in E(n)$. Then define

$$\text{Ad}(A)(Y) = [A, Y] \equiv AY - YA, \quad \forall Y \in E(n).$$

Clearly Ad(A) may be viewed as a linear map of $E(n)$ into E. (Most of these algebraic ideas may be found discussed in Chevalley, 1946.)

Since $GL \subseteq E$, it is meaningful to consider the mapping exp: $E \to GL$. Thus, we may state the following theorem connecting the three maps Ad, ad and exp.

THEOREM 6.2.1. *Let $A \in E(n)$, then* $\exp(\text{Ad}(A)) = \text{ad}(\exp(a))$.
Proof. The proof of this theorem is by computation using the definition of the three mappings exp, ad and Ad.   □

The subspace S
Let $B \in E(n)$. The quantity ad($\exp(At)$)B is what is being averaged in (6.1.53) to determine $\bar{B}$. Thus we are led to describe the span of ad($\exp(At)$)B as t ranges over R. We will see that this span is a subspace $S = S(B)$ of E.

Let $V_0 = B$ and $V_k = \text{Ad}(A)V_{k-1}, k = 1, 2, \ldots$. Let N be the first index such that $V_0, \ldots, V_{N+1}$ are linearly dependent, and let

$$V_{N+1} = \sum_{i=0}^{N} a_i V_i.$$

Let S be the subspace of E spanned by $\{V_i | i = 0, \ldots, N\}$.

The following lemma connects S with the averages of interest.

LEMMA 6.2.2. *Each of the following two sets span S.*

(i) $\left\{ \dfrac{d^k}{dt^k} \text{ad}(\exp(tA))B\big|_{t=0} \big| k = 0, 1, \ldots \right\}$

(ii) $\{\text{ad}(\exp(tA))B | \ t \in R\}$.

Proof. (i) easily follows since

$$\frac{d^k}{dt^k}\operatorname{ad}(\exp(tA))B\Big|_{t=0} = \operatorname{Ad}^k(A)(B).$$

To show (ii), we begin by noting that S is $\operatorname{Ad}(tA)$ invariant so that it is $\exp(\operatorname{Ad}(tA))$ invariant as well. Then Theorem 6.2.1 implies that $\exp(\operatorname{Ad}(tA)) = \operatorname{ad}(\exp(tA))$. Thus, S is $\operatorname{ad}(\exp(tA))$ invariant. Thus since $B \in S$, then $\exp(\operatorname{Ad}(tA))(B) \in S$, $\forall t$. Now in fact, part (i) shows that $\exp(\operatorname{Ad}(tA))(B)$ spans S for t close to $t = 0$; a fortiori it spans S for all t. This completes the proof of the lemma. $\qquad\square$

By construction, S is a cyclic subspace for $\operatorname{Ad}(A)$ with cyclic vector B. Thus the restriction of $\operatorname{Ad}(A)$ to S has the matrix representation

$$\begin{bmatrix} 0 & 1 & 0 & . & . & . & & 0 \\ 0 & 0 & 1 & 0 & . & . & . & 0 \\ & & . & . & . & & & \\ 0 & & . & . & . & & 0 & 1 \\ a_0 & & . & . & . & & & a_N \end{bmatrix},$$

relative to the base $V_0, \ldots, V_N$ of S. Thus the determinant of $\operatorname{Ad}(A)$ is a_0. The following theorem characterizes averaging algebraically.

THEOREM 6.2.3. *Let the eigenvalues of A be purely imaginary. Let*

$$\bar{B} = \lim_{T \to \infty} \frac{1}{T} \int_0^T \operatorname{ad}(\exp(tA))B\,dt,$$

and let $V_0, \ldots, V_{n+1} = \sum_{i=0}^{N} a_i V_i$. Let S be as defined above.

(i) *If $a_0 \neq 0$, then $\bar{B} = 0$.*

(ii) *If $a_0 = 0$, there exists a non-zero $W \in S$ which is unique up to a scalar multiple such that $\operatorname{Ad}(A)W = 0$. If $W = \sum_{i=0}^{N} b_i V_i$, then $\bar{B} = -W/b_0$.* $\qquad\square$

An Invariant Statement of Theorem 6.2.3

Before giving the proof of Theorem 6.2.3, we rephrase this result in an *invariant terminology* which frees it from the coordinate system $V_0, \ldots, V_N$ used in the statement of this theorem.

According to Theorem 6.2.3 if $\operatorname{Ad}(A)$ is nonsingular, then $\bar{B} = 0$, while if Ad(A) is singular, then $\bar{B} = -W/b_0$. Since $W = \sum_{i=0}^{N} b_i V_i$ and $B = V_0$, we have

$$B = -\frac{W}{b_0} + \frac{1}{b_0} \sum_{i=1}^{N} b_i V_i. \tag{6.2.1}$$

The uniqueness of W implies that the first term in this sum is a vector in the one dimensional eigenvector space $S_0 \subseteq S$, corresponding to the eigenvalue zero of Ad(A). By definition of the V_i, the second term in (6.2.1) lies in the subspace $S_1 \subseteq S$, which is the range of Ad(A) (restricted to S). Since $S = S_0 \oplus S_1$, if $B = B_0 + B_1$ with $B_0 \in S_0$ and $B_1 \in S_1$, then $\bar{B} = B_0$. That is to say, $\bar{B}$ is the projection along the range of Ad(A) (restricted to S) onto the one dimensional null space of Ad(A) (restricted to S).

This invariant statement of the main result depends on S and the construction of its base $V_0, \ldots, V_N$. In fact, even this dependence can be eliminated.

To see this, we begin by noting that Ad(A) is a *completely reducible transformation* acting on $E(n)$. Thus

$$E(n) = E_0(n) \oplus E_1(n),$$

where $E_0(n)$ is the null space of Ad(A) and $E_1(n)$ is its range.

Clearly $S_0 \subseteq E_0(n)$ and $S_1 \subseteq E_1(n)$. Thus since $B = B_0 + B_1$ with $B_0 \in S_0$ and $B_1 \in S_1$, we have $B_0 \in E_0(n)$ and $B_1 \in E_1(n)$. Thus $B = B_0 + B_1$ is also a decomposition in B into a part in $E_0(n)$ and a part in $E_1(n)$. However, any such decomposition is well-defined. Thus B_0 is the projection along $E_1(n)$ onto $E_0(n)$. If P denotes this projection operator, we have that

$$\bar{B} = P(B) = \lim_{T \to \infty} \frac{1}{T} \int_0^T e^{At} B e^{-At} dt.$$

In Section 6.2.2, we show how to carry out the computation $P(B)$ when the matrix A has a particularly simple form.

We make use of the following lemma in the proof of Theorem 6.2.3.

LEMMA 6.2.4. *If A has purely imaginary eigenvalues, then so does* Ad(tA). *Proof.* By hypothesis $\exp(tA)$ is contained in the orthogonal group (which is compact). Then C, the closure of $\exp(tA)$ (in the orthogonal group) is compact. Then ad(C) is a compact group operating as linear transformations on $E(n)$. But any compact group is similar to a subgroup of the orthogonal group on $E(n)$. (See Hochshild, 1965.) Thus every matrix in ad(C) is similar to a diagonal matrix whose jjth entry is $\exp(ik_j t)$ for some $k_j \in R$.

Using Theorem 6.2.1, we have

$$\text{Ad}(A) = \frac{d}{dt}\exp\left(\text{Ad}(tA)\right)\Big|_{t=0} = \frac{d}{dt}\,\text{ad}(\exp(tA))\Big|_{t=0}.$$

That is, $\text{Ad}(A)$ is similar to a diagonal matrix whose jjth entry is $(d/dt)\exp(ik_j t)\big|_{t=0} = ik_j$. This completes the proof of the lemma. $\qquad\square$

We now give the proof of Theorem 6.2.3, and in particular, in its invariant terminology form.

Proof of Theorem 6.2.3. Lemma 6.2.4 shows that there exists a coordinate system in S relative to which $\text{Ad}(tA)$ has the following matrix representation.

$$\text{Ad}(tA) = \begin{bmatrix} i\lambda_0 t & & & & 0 \\ & \cdot & & & \\ & & \cdot & & \\ & & & \cdot & \\ 0 & & & & i\lambda_N t \end{bmatrix}.$$

Let B have the coordinates $[\beta_0, \ldots, \beta_N]^T$ in this coordinate system. Then using Theorem 6.2.1,

$$\exp(\text{Ad}(tA))B = \text{ad}(\exp(tA))B = \begin{bmatrix} e^{i\lambda_0 t}\beta_0 \\ \cdot \\ \cdot \\ \cdot \\ e^{i\lambda_N t}\beta_N \end{bmatrix}.$$

This and part (i) of Lemma 6.2.2 imply that the following vectors,

$$\begin{bmatrix} \beta_0 \\ \cdot \\ \cdot \\ \cdot \\ \beta_N \end{bmatrix}, \begin{bmatrix} \lambda_0\beta_0 \\ \cdot \\ \cdot \\ \cdot \\ \lambda_N\beta_N \end{bmatrix}, \ldots, \begin{bmatrix} \lambda_0^N\beta_0 \\ \cdot \\ \cdot \\ \cdot \\ \lambda_N^N\beta_N \end{bmatrix},$$

span S. Thus the matrix whose columns are these vectors is nonsingular. This matrix may be written as the following product of a diagonal matrix and a Vandermonde matrix.

$$\begin{bmatrix} \beta_0 & & & 0 \\ & \cdot & & \\ & & \cdot & \\ & & & \cdot \\ 0 & & & \beta_N \end{bmatrix} \begin{bmatrix} 1 & \lambda_0 & \cdot & \cdot & \cdot & \lambda_0^N \\ \cdot & \cdot & & & & \cdot \\ \cdot & & \cdot & & & \cdot \\ \cdot & & & \cdot & & \cdot \\ 1 & \lambda_N & \cdot & \cdot & \cdot & \lambda_N^N \end{bmatrix}.$$

The nonsingularity of this matrix product implies that no β_j vanishes and that the λ_j are distinct.

Evaluating $\bar{B}$ amounts to evaluating

$$\lim_{T \to \infty} \frac{1}{T} \int_0^T e^{i\lambda_j t} \beta_j \, dt, \quad j = 0, ., N.$$

This limit is zero if $\lambda_j \neq 0$ and it equals β_j if $\lambda_j = 0, j = 0, \ldots, N$.

If no $\lambda_j = 0$, then $\mathrm{Ad}(A)$ is nonsingular, and the proof is complete. Suppose then that $\mathrm{Ad}(A)$ is singular. Since the λ_j are distinct, at most one λ_j can vanish. Thus the space S_0 annihilated by $\mathrm{Ad}(A)$ has dimension one. By relabelling, we may suppose that $\lambda_0 = 0$ so that the range S_1 of $\mathrm{Ad}(A)$ consists of vectors of the form $[0, y_1, \ldots, y_N]^T$. Thus writing $B = B_0 + B_1$ with $B_0 \in S_0$ and $B_1 \in S_1$, we have

$$B_0 = \begin{bmatrix} \beta_0 \\ 0 \\ \cdot \\ \cdot \\ \cdot \\ 0 \end{bmatrix} \quad \text{and} \quad B_1 = \begin{bmatrix} 0 \\ \beta_1 \\ \cdot \\ \cdot \\ \cdot \\ \beta_N \end{bmatrix}.$$

Thus $\bar{B} = \beta_0$, completing the proof of the theorem. $\qquad\square$

6.2.2. An Example

Let us apply the theory just developed to the differential equation describing a canonical mechanical system:

$$M\ddot{q} + Cq = 0.$$

q is the position vector, M is an inertial matrix and C is the stiffness matrix.

M and C are symmetric and positive definite. We introduce matrices L and R through the following relations.

$$M = L^{-1}R^{-1}, \quad LC = R^T$$

Then introducing a momentum vector p, the differential equation becomes

$$\begin{bmatrix} \dot{q} \\ \dot{p} \end{bmatrix} = \begin{bmatrix} 0 & R \\ -R^T & 0 \end{bmatrix} \begin{bmatrix} q \\ p \end{bmatrix}.$$

Typically $M = I$, the identity matrix, so that we may take

$$L^{-1} = C^{1/2}, R^{-1} = C^{1/2} \quad and \quad R^T = LC = C^{1/2}.$$

Setting $a = C^{1/2}$, so that a is symmetric and nonsingular, the differential equation becomes

$$\begin{bmatrix} \dot{q} \\ \dot{p} \end{bmatrix} = \begin{bmatrix} 0 & a \\ -a & 0 \end{bmatrix} \begin{bmatrix} q \\ p \end{bmatrix}.$$

Then for this example,

$$A = \begin{bmatrix} 0 & a \\ -a & 0 \end{bmatrix}.$$

Let B be the following matrix.

$$B = \begin{bmatrix} \alpha & \beta \\ \gamma & \delta \end{bmatrix},$$

where α, β, λ and δ are blocks corresponding to the blocks of A. Let $V_0 = B$ be taken in the following form.

$$V_0 = \begin{bmatrix} \alpha \\ \delta \\ \beta \\ \gamma \end{bmatrix}$$

Further introduce the matrices E, F and G:

$$E = \begin{bmatrix} I & I \\ I & I \end{bmatrix}, \quad F = \begin{bmatrix} I & -I \\ -I & I \end{bmatrix}, \quad G = \begin{bmatrix} I & I \\ -I & -I \end{bmatrix}, \quad (6.2.2)$$

where the blocks here also correspond to the blocks of A. Then we find that for $n = 1, 2, \ldots,$

$$(\mathrm{Ad}(A))^n = 2^{n-1}a^n \begin{cases} (-1)^{(n-1)/2}\begin{bmatrix} 0 & G \\ -G^T & 0 \end{bmatrix}, & n \text{ odd,} \\[2em] (-1)^{n/2}\begin{bmatrix} F & 0 \\ 0 & E \end{bmatrix}, & n \text{ even.} \end{cases}$$

Here and in what follows, this notation means that the matrix a multiplies each of the blocks in the other indicated matrices. (In this equation there are 16 such blocks per matrix. See (6.2.2).)

Then

$$\mathrm{Range\ Ad}(A) = \left\{ V \,\middle|\, V = a\begin{bmatrix} 0 & G \\ -G^T & 0 \end{bmatrix}\begin{bmatrix} \alpha \\ \delta \\ \beta \\ \gamma \end{bmatrix} \right\}$$

$$= \left\{ V \,\middle|\, V = a\begin{bmatrix} \beta + \gamma \\ -\beta - \gamma \\ -\alpha + \delta \\ -\alpha + \delta \end{bmatrix} \right\}.$$

Thus, if x and w are blocks of the size of a, we have

$$\mathrm{Range\ Ad}(A) = \left\{ V \,\middle|\, V = a\begin{bmatrix} w \\ -w \\ x \\ x \end{bmatrix} \right\}.$$

We similarly find that if u and v are blocks of size a, that

$$\mathrm{Null\ Ad}(A) = \left\{ a\begin{bmatrix} u \\ u \\ v \\ -v \end{bmatrix} \right\}.$$

Thus writing B as a sum of vectors in the null space and range respectively, requires a representation of the following form.

$$\begin{bmatrix} \alpha & \beta \\ \gamma & \delta \end{bmatrix} = a\begin{bmatrix} u & v \\ -v & u \end{bmatrix} + a\begin{bmatrix} w & x \\ x & -w \end{bmatrix}.$$

Thus,

$$u = \tfrac{1}{2}a^{-1}(\alpha + \delta),$$
$$v = \tfrac{1}{2}a^{-1}(\beta - \gamma),$$
$$w = \tfrac{1}{2}a^{-1}(\alpha - \delta),$$
$$x = \tfrac{1}{2}a^{-1}(\beta + \gamma).$$

Thus for the projection $P(B)(= \bar{B})$ of B along the range of Ad(A) and onto the null space of Ad(A), we find

$$P(B) = \frac{1}{2}\begin{bmatrix} \alpha + \delta & \beta - \gamma \\ \gamma - \beta & \alpha + \delta \end{bmatrix}.$$

6.2.3. Preconditioning

The discussion concerning the invariant statement of Theorem 6.2.3 in Section 6.2.1 shows that the average $\bar{B} = P(B)$, the projection along the range of Ad(A) onto the null space of Ad(A). We may carry out this computation when A has a particularly simple form.

First suppose that A is the 2×2 matrix J, where

$$J = \begin{bmatrix} 0 & 1 \\ -1 & 0 \end{bmatrix}.$$

Then we have the following lemma whose proof follows by direct computation.

LEMMA 6.2.5. Ad(J) is defined on $E(2) = \text{Null}(\text{Ad}(J)) \oplus \text{Range }(\text{Ad}(J))$, where

$$\text{Null}(\text{Ad}(J)) = \begin{bmatrix} a & b \\ -b & a \end{bmatrix}$$

and

$$\text{Range}(\text{Ad}(J)) = \begin{bmatrix} c & d \\ d & -c \end{bmatrix}.$$

Thus, given the arbitrary matrix $\begin{bmatrix} \alpha & \beta \\ \gamma & \delta \end{bmatrix}$, we compute

$$a = \frac{\alpha + \delta}{2}, \quad b = \frac{\beta - \gamma}{2}, \quad c = \frac{\alpha - \delta}{2}, \quad d = \frac{\beta + \gamma}{2}.$$

Then

$$\begin{bmatrix} \alpha & \beta \\ \gamma & \delta \end{bmatrix} = \begin{bmatrix} a & b \\ -b & a \end{bmatrix} + \begin{bmatrix} c & d \\ d & -c \end{bmatrix},$$

and

$$P\begin{bmatrix} \alpha & \beta \\ \gamma & \delta \end{bmatrix} = \begin{bmatrix} \dfrac{\alpha + \delta}{2} & \dfrac{\beta - \gamma}{2} \\ \dfrac{\gamma - \beta}{2} & \dfrac{\alpha + \delta}{2} \end{bmatrix}.$$

Let J_s be the block diagonal matrix with s blocks of J along the main diagonal. Then for an arbitrary skew symmetric matrix of even order, we have the following lemma.

LEMMA 6.2.6. *An even ordered skew symmetric matrix is similar to a block diagonal matrix with the blocks $k_i J_{s_i}$, $i = 1, \ldots, N$ along the main diagonal.*

(We suppose that the skew symmetric matrix is nonsingular, so that no k_i vanishes. Moreover since the eigenvalues of J are $\pm\, i$, we assume without loss of generality that $k_i > 0, i = 1, \ldots, N$.)

The following observation describes what is meant by preconditioning here.

REMARK 6.2.7. The similarity transformation $R^T AR$, where R is the orthogonal matrix whose columns are the real parts and the imaginary parts of the eigenvalues of A, will produce the block diagonal form referred to in Lemma 6.2.6. (See Bellman, 1960, p. 64.) Determining R may be viewed as a preconditioning of the averaging problem when there are many matrices B to be averaged (as is the case of solving nonlinear differential equations at many mesh points). Indeed with this preconditioning having been performed, the averaging process becomes quite simple, as we will presently see.

Suppose that A has the block diagonal form described in Lemma 6.2.6. Let any given matrix B of order $\sum\limits_{i=1}^{N} s_i$, (i.e., the order of A, the skew symmetric matrix under consideration) be viewed as a matrix of blocks B_{ij}, $i, j = 1, \ldots, N$, corresponding to the blocks of A. Thus B_{ij} is a submatrix of order $2s_i \times 2s_j$, $i, j = 1, \ldots, N$. Each such block corresponds to a subspace of $E(n)$ which is invariant under $\text{Ad}(A)$. To see this, let $\hat{B}_{ij}$ be the matrix obtained from B by setting every element in B equal to zero except for those in the one block B_{ij}, $i, j = 1, \ldots, N$. Then in particular,

$$B = \sum_{i,j=1}^{N} \hat{B}_{ij}.$$

Now a computation shows that $\text{Ad}(A)\, \hat{B}_{ij}$ is zero everywhere except in the ijth block, where it has the value

$$k_i J_{s_i} B_{ij} - k_j B_{ij} J_{s_j}.$$

Now let each block B_{ij} be composed of the 2×2 subblocks C_{ij}^{lm}, $l = 1, \ldots, s_i, m = 1, \ldots, s_j$, follows.

$$B_{ij} = \begin{bmatrix} C_{11}^{ij} & \cdots & C_{1s_j}^{ij} \\ \cdot & & \cdot \\ \cdot & & \cdot \\ C_{s_i 1}^{ij} & \cdots & C_{s_i s_j}^{ij} \end{bmatrix}.$$

Then the lmth 2×2 block of $k_i J_{s_j} B_{ij} - k_j B_{ij} J_{s_j}$ is

$$k_i J C_{lm}^{ij} - k_j C_{lm}^{ij} J, \quad \forall i, j, l, m. \tag{6.2.3}$$

In particular if $\hat{B}_{ij;lm}$ is the matrix obtained from $\hat{B}_{ij}$ by setting every element equal to zero except the lmth 2×2 subblock of B_{ij} itself, then

$$B = \sum_{i,j=1}^{N} \sum_{l=1}^{s_i} \sum_{m=1}^{s_j} \hat{B}_{ij;lm}.$$

Moreover, a computation shows that $\text{Ad}(A)\hat{B}_{ij;lm}$ is everywhere zero except in lmth 2×2 subblock of the subblock B_{ij}, where it has the value (6.2.3). That is, every 2×2 subblock C_{ij}^{lm} of B corresponds to an invariant subspace of $\text{Ad}(A)$.

Thus to find $\text{Ad}(A)B$ it is only necessary to compute $\text{Ad}(A)\hat{B}_{ij,lm}$ (see (6.2.3)) and add.

To find $P(B)$ the projection of B along the range of $\text{Ad}(A)$ onto the null space of $\text{Ad}(A)$, we may use the invariant subspaces, as just noted, and find this projection for each such subspace in turn. In fact, there are only two simple possibilities.

If $i = j$ so that B_{ij} is a diagonal block, $P(\hat{B}_{ii;lm})$ is everywhere zero except in the nonvanishing 2×2 subblock to which $\hat{B}_{ii;lm}$ corresponds, and where from Lemma 6.2.5, we see that $P(\hat{B}_{ii;lm})$ has the value

$$
\begin{bmatrix}
\dfrac{C^{ii}_{lm_11} + C^{ii}_{lm_122}}{2} & \dfrac{C^{ii}_{lm_112} - C^{ii}_{lm;21}}{2} \\[4mm]
\dfrac{C^{ii}_{lm;21} - C^{ii}_{lm_112}}{2} & \dfrac{C^{ii}_{lm;11} + C^{ii}_{lm;22}}{2}
\end{bmatrix}.
$$

Here $C^{lm}_{ij;pq}$, $p,q = 1,2$, are the four components of the 2×2 submatrix C^{lm}_{ij}. If $i \neq j$, we seek the null space of the transformation

$$
\mathrm{Ad}(A)\hat{B}_{ij;lm} = k_i J C^{ij}_{lm} - k_j C^{ij}_{lm} J, \quad \forall i \neq j.
$$

Setting this to zero gives the following matrix equation.

$$
k_i \begin{bmatrix} C_{21} & C_{22} \\ -C_{11} & -C_{12} \end{bmatrix} = k_j \begin{bmatrix} -C_{12} & C_{11} \\ -C_{22} & C_{21} \end{bmatrix}, \quad \forall i \neq j.
$$

Here we have suppressed the index pairs ij and lm for clarity.

These equations yield $k_i = \mp k_j$. Since $k_i > 0$, $i = 1, \ldots, N$, the null space we seek is empty. Thus

$$
P(\hat{B}_{ij;lm}) = 0.
$$

We summarize this discussion as follows.

Summary
Suppose A has the simple form specified in Lemma 6.2.6. Then B may be determined by considering the block structure given by A. In the off-diagonal blocks, B vanishes. In the diagonal blocks, simply apply Lemma 6.2.5 to each 2×2 subblock. See Auslander and Miranker, 1979 for additional details.

6.3. ACCELERATED COMPUTATION OF AVERAGES AND AN EXTRAPOLATION METHOD

In this section, we consider techniques for accelerating the computation of averages. We also develop extrapolation methods which replace the stiff highly oscillatory problem with auxiliary relaxed equations. We begin with a review of the multi-time expansion in the nonlinear case.

6.3.1. *The Multi-time Expansion in the Nonlinear Case*

Consider the following nonlinear analogue of the model problem (6.1.1) which we have been studying.

$$dx/dt = f(t/\varepsilon, x), \quad x(0) = \xi, \tag{6.3.1}$$

where $x, f, \xi \in R^n$ and where $f(\tau, \cdot)$ is an almost periodic function of τ. Multi-time perturbation methods lead to the approximation

$$x(t, \varepsilon) = x_0(t) + \varepsilon x_1(t, t/\varepsilon) + O(\varepsilon^2), \tag{6.3.2}$$

(the analogue of (6.1.15)), where x_0 is determined from the initial value problem

$$dx_0/dt = \bar{f}(x_0), \quad x_0(0) = \xi, \tag{6.3.3}$$

(the analogue of (6.1.25). See (6.1.52) also. Also see Volosov, 1962.)
    As usual, $\bar{f}$ is the average of f, defined by

$$\bar{f}(x_0) = \lim_{T \to \infty} \frac{1}{T} \int_0^T f(\tau, x_0) d\tau,$$

(the analogue of (6.1.24)).
The coefficient x_1 is determined from the formula

$$x_1(t, t/\varepsilon) = \tilde{x}_1(t) + \int_0^{t/\varepsilon} [f(\tau, x_0) - \bar{f}(x_0)] d\tau. \tag{6.3.4}$$

In this formula, $\tilde{x}_1$ is determined at a later step in the perturbation scheme. Since it will not be needed here, it is not discussed further. (For such details, see Persek and Hoppensteadt, 1978.) Thus,

$$x(t, \varepsilon) = x_0(t) + \varepsilon \left\{ \tilde{x}_1(t) + \int_0^{t/\varepsilon} [f(\tau, x_0) - \bar{f}(x_0)] d\tau \right\} + O(\varepsilon^2).$$

    This approximation suggests several numerical schemes for determining $x(h, \varepsilon)$. In Section 6.3.2, we consider the computation of the average $\bar{f}$; first by the customary method and then by a second difference method which accelerates the computation of $\bar{f}$ in some cases. Then in Section 6.3.3, we describe an extrapolation method for approximating $x(h, \varepsilon)$. As in the extrapolation method introduced in Section 5.3, certain larger values ε' of ε are introduced, and (6.3.1) is used with this larger value of ε' to furnish approximations to $x(h, \varepsilon)$ itself. In Section 6.3.4, the results of computational experiments which compare the methods are presented. Finally a discussion of these various computational procedures is given in Section 6.3.5.

6.3.2. *Accelerated Computation of* $\bar{f}$

We propose two methods for calculating $\bar{f}$.

(i) *Direct evaluation of* $\lim\limits_{T \to \infty} \dfrac{1}{T} \displaystyle\int_0^T f \, dt$.

A convergence criterion is first set, and then the integral $\dfrac{1}{T} \displaystyle\int_0^T f \, d\tau$ is calculated for increasing values of T until the criterion is met: Given a tolerance δ. there is a value $T(\delta, x)$ such that

$$\left\| \frac{1}{T_1} \int_0^{T_1} f(\tau, x) d\tau - \frac{1}{T_2} \int_0^{T_2} f(\tau, x) d\tau \right\| < \delta$$

and

$$\left\| \frac{1}{T_1} \int_0^{T_1} f(\tau, x) d\tau - \bar{f}(x) \right\| < \delta$$

for all $T_1, T_2 \geq T(\delta, x)$. Thus, we can write

$$\bar{f}(x) = \frac{1}{T(\delta, x)} \int_0^{T(\delta, x)} f(\tau, x) d\tau + O(\delta).$$

and proceed to solve (6.3.3). Unfortunately, there is no certain way of finding $T(\delta, x)$.

In order to find a candidate for $T(\delta, x)$, we calculate

$$V(T, x) = \int_0^T f(\tau, x) d\tau$$

for $0 \leq T \leq 2T^*$, and keep increasing T^* until the condition

$$\sup_{0 \leq T' \leq T} \left\| \frac{1}{T^*} V(T^*, x) - \frac{1}{T^* + T'} V(T^* + T', x) \right\| \leq \delta$$

is met. Then we take $T(\delta, x) = 2T^*$. Usually $2T^*$ is of order $1/\delta$. For example, if $f(\tau, x) = \sin \tau$, then $\bar{f}(x) = 0$ and $V(T, x) = \cos T - 1$. Thus

$$0 \leq \left\| \frac{1}{T} V(T, x) \right\| \leq 2/T,$$

the maximum being attained in each interval of length 2π.

(ii) *Second difference method*

In most applications, the integral of the almost periodic function f has the form

$$V(T, x) = \bar{f}(x)T + p(T, x),$$

where p is an almost periodic function of its first argument. Thus, given a tolerance δ, there is a δ-translation number $\mathscr{T}(\delta, x)$ such that

$$\| p(T + \mathscr{T}(\delta, x), x) - p(T, x) \| < \delta$$

for all $T \geq 0$; in particular, since $p(0, x) = 0$, then $|p(\mathscr{T}(\delta, x)| < \delta$.

If the frequencies of V, hence p, are known, then $\mathscr{T}(\delta, x)$ can be determined by a Diophantine approximation procedure (see Leveque, 1956). Thus, Fourier transform methods can be used to determine the spectrum, and a Diophantine algorithm used to find $\mathscr{T}(\delta, x)$. We do not carry this procedure out here. Instead we find candidates for $\mathscr{T}$ by an alternate method.

Note that

$$V(2T, x) - 2V(T, x) = p(2T, x) - 2p(T, x).$$

In particular for $T = \mathscr{T}(\delta, x)$, we have that

$$
\begin{aligned}
V(2\mathscr{T}, x) - 2V(\mathscr{T}, x) &= p(2\mathscr{T}, x) - p(\mathscr{T}, x) - p(\mathscr{T}, x) + p(0, x) \\
&= O(\delta).
\end{aligned}
\tag{6.3.5}
$$

Thus, any δ-translation number p makes (6.3.5) of order δ. Unfortunately the converse does not hold; in particular, $\| V(2T, x) - 2V(T, x) \|$ may be small while $\| p(T, x) \|$ is not small.

Still, by tabulating

$$V(2T, x) - 2V(T, x),$$

candidates for $\mathscr{T}(\delta, x)$ can be found and tested by comparing the values of $V(T, x)/T$ for several of them, since these should all approximate $\bar{f}(x)$. In practice, this method is no worse than the direct calculation in (i), and in periodic cases, it reliably gives $\bar{f}(x)$ after calculation over one period.

Thus from either (i) or (ii), we use

$$\frac{1}{\mathscr{T}(\delta, x)} \int_0^{\mathscr{T}(\delta, x)} f(\tau, x) d\tau$$

as an approximation to $\bar{f}$ and proceed to integrate (6.3.1) using this approximation.

6.3.3. *The Extrapolation Method*

As in the matched asymptotic expansions case (see Section 5.4), an appropriate value T must be found which represents a time at which rapid motions can be ignored. In the present case, we pick T to be a δ-translation number of $p(\tau, x)$. Then, in particular,

$$\frac{p(2T, x) - 2p(T, x)}{T} = \frac{1}{T} \int_T^{2T} [f(\tau, x) - \bar{f}(x)] d\tau = O(h^p), \qquad (6.3.6)$$

for $x = \xi + O(h)$. The existence of such a value of T follows from viewing (6.3.6) as the statement that T is an approximation to a δ-translation number (compare (6.3.5) with $O(\delta) = h^p$). Such a T value must be found, perhaps using one of the methods mentioned earlier in this section or additional knowledge about a specific problem being studied.

Once a T value is found which satisfies (6.3.6), we define

$$\varepsilon' = h/T$$

as in Section 5.3, and then we calculate $x(h, \varepsilon'/2)$ and $x(h, \varepsilon')$ from (6.3.1) by a pth-order numerical method. It follows from formulas (6.3.2) and (6.3.4) that

$$2x(h, \varepsilon'/2) - x(h, \varepsilon) = x_0(h) + \varepsilon' \int_T^{2T} [f(\tau, x_0(h)) - \bar{f}(x_0(h))] d\tau$$
$$+ O((\varepsilon')^2),$$
$$= x_0(h) + O(h^{p+1}) + O((h/T)^2).$$

On the other hand,

$$x(h, \varepsilon) = x_0(h) + \varepsilon \tilde{x}_1 + \varepsilon \int_0^{h/\varepsilon} [f(\tau, x_0(h)) - \bar{f}(x_0(h))] d\tau + O(\varepsilon^2)$$
$$= x_0(h) + O(\varepsilon),$$

since

$$\int_0^{h/\varepsilon} [f(\tau, x_0(h)) - \bar{f}(x_0(h))] d\tau = O(1).$$

Therefore,

$$x(h, \varepsilon) = 2x(h, \varepsilon'/2) - x(h, \varepsilon') + O(\varepsilon) + O(h^{p+1}) + O((h/T)^2). \quad (6.3.7)$$

This formula gives the extrapolation method for calculating $x(h, \varepsilon)$. It and the previous methods are compared for an example in the next section.

We refer here to Remark 5.3.1 to emphasize that the approach represented by the extrapolation method is important both computationally and theoretically. (See Remark 5.3.1a especially for the latter, where the role of the extrapolation method in avoiding straightforward evaluation of terms of an expansion is explained.)

Note that if T differs by the quantity Δ from the approximation $\mathcal{T}(\delta, x)$ of the δ-translation number, then from (6.3.7) we see that the corresponding difference in the associated value of $x(h, \varepsilon)$ is

$$O(h^2 / T^3) .$$

The dependence of this estimate on h is illustrated in Table 6.3-1.

6.3.4. Computational Experiments: A Linear System

The linear initial value problem

$$\frac{du}{dt} = ((1/\varepsilon)A + B)u, \qquad u(0) = \xi,$$

is taken into the problem

$$\frac{dv}{dt} = e^{-At/\varepsilon} B e^{At/\varepsilon} v, \qquad v(0) = \xi$$

by the change of variables $u = e^{At/\varepsilon} v$. We take $v \in R^4$ and $\xi = (1, 1, 1, 1)^T$.

Running values of $V(T, \xi)$ are computed using a quadrature increment of $\Delta T = 0.01$. The tolerances for each of the methods (i) and (ii) are denoted by δ_i and δ_{ii}, respectively, and the corresponding values of T at which the associated computations halt are denoted T_i and T_{ii}, respectively. The calculations are carried out for two different matrices:

$$A(1, w) = \begin{bmatrix} 0 & 1 & 0 & 0 \\ -1 & 0 & 0 & 0 \\ 0 & 0 & 0 & w \\ 0 & 0 & -w & 0 \end{bmatrix}, \quad A(2, w) = \begin{bmatrix} 0 & 1 & 2 & 0 \\ -1 & 0 & 0 & 3 \\ 0 & 0 & 0 & w \\ 0 & 0 & -w & 0 \end{bmatrix},$$

where w is a parameter specified below.

B is taken to be

$$B = \begin{bmatrix} 1 & 2 & 3 & 4 \\ 5 & 6 & 7 & 8 \\ 8 & 7 & 6 & 5 \\ 4 & 3 & 2 & 1 \end{bmatrix}.$$

TABLE 6.3-1: Case 1. $A = A(1, \frac{1}{2})$

(a) Approximations to $\bar{B}$

		$\bar{B}(1,\frac{1}{2})$			$\delta_i = 0.001$ $\;\;\;T_i = 163.36$	$\bar{B}_i(1,\frac{1}{2})$			$\delta_{ii} = 0.001$ $\;\;\;T_{ii} = 25.12$	$\bar{B}_{ii}(1,\frac{1}{2})$		
3.5	−1.5	0.0	0.0	3.5	−1.501	−0.001	0.0	3.501	−1.502	−0.002	−0.002
1.5	3.5	0.0	0.0	1.5	3.5	0.0	0.0	1.498	3.499	−0.004	−0.004
0.0	0.0	3.5	−1.5	0.0	0.0	3.5	−1.5	−0.004	−0.004	3.499	1.498
0.0	0.0	1.5	3.5	0.0	0.0	1.5	3.5	−0.002	−0.002	−1.502	3.501

(b) Approximations to $x(h, \varepsilon)$, $h = 0.1$

Approximation	Approximating vector			
$e^{\bar{B}h}\zeta$	1.192	1.614	1.614	1.192
$e^{\bar{B}_i h}\zeta$	1.197	1.602	1.602	1.197
$e^{\bar{B}_{ii}h}\zeta$	1.197	1.601	1.601	1.197
$Ex(0.0006)$	1.193	1.615	1.616	1.191
$Ex(0.004)$	1.194	1.618	1.614	1.193

(c) Approximations to $x(h, \varepsilon)$, $h = 0.05$

Approximation	Approximating vector			
$e^{\bar{B}h}\zeta$	1.099	1.277	1.277	1.099
$e^{\bar{B}_i h}\zeta$	1.099	1.276	1.276	1.099
$e^{\bar{B}_{ii}h}\zeta$	1.099	1.275	1.275	1.099
$Ex(0.003)$	1.099	1.277	1.277	1.098
$Ex(0.002)$	1.099	1.277	1.277	1.098

TABLE 6.3.1: Case 2. $A = A(1, \frac{1}{4})$

(a) Approximations to $\bar{B}$

$\bar{B}(1,\tfrac{1}{4})$				$\bar{B}_i(1,\tfrac{1}{4})$ — $\delta_i = 0.001$, $T_i = 158.34$				$\bar{B}_{ii}(1,\tfrac{1}{4})$ — $\delta_{ii} = 0.001$, $T_{ii} = 50.28$			
3.5	-1.5	0.0	0.0	3.475	-1.508	-0.085	-0.027	3.499	-1.499	0.001	0.001
1.5	3.5	0.0	0.0	1.402	3.525	-0.024	0.026	1.501	3.501	0.002	0.002
0.0	0.0	3.5	1.5	-0.085	0.0265	3.401	1.531	0.002	0.002	3.501	3.501
0.0	0.0	-1.5	3.5	0.024	0.261	-1.469	3.599	0.001	0.001	0.002	3.499

(b) Approximations to $x(h, \varepsilon), h = 0.1$

Approximation	Approximating vector			
$e^{\bar{B}h}\zeta$	1.192	1.614	1.614	1.192
$e^{\bar{B}_i h}\zeta$	1.178	1.604	1.587	1.222
$e^{\bar{B}_{ii}h}\zeta$	1.198	1.603	1.603	1.198
$Ex(0.0006)$	1.199	1.607	1.638	1.168
$Ex(0.002)$	1.193	1.617	1.617	1.192

(c) Approximations to $x(h, \varepsilon), h = 0.05$

Approximation	Approximating vector			
$e^{\bar{B}h}\zeta$	1.099	1.277	1.277	1.099
$e^{\bar{B}_i h}\zeta$	1.091	1.276	1.269	1.110
$e^{\bar{B}_{ii}h}\zeta$	1.100	1.276	1.276	1.100
$Ex(0.0003)$	1.102	1.273	1.287	1.088
$Ex(0.001)$	1.102	1.273	1.287	1.088

TABLE 6.3-1: Case 3. $A = A(2, \frac{1}{2})$

(a) Approximations to $\bar{B}$

$\bar{B}(2, \frac{1}{2})$				$\bar{B}_i(2, \frac{1}{2})$ $\delta_i = 0.01$, $T_i = 404.28$				$\bar{B}_{ii}(2, \frac{1}{2})$ $\delta_{ii} = 0.0001$, $T_{ii} = 75.4$			
9.167	-28.167	-290.556	-60.484	9.104	-28.028	-289.046	-60.117	9.166	-28.166	-290.599	-60.443
28.167	-9.167	-52.889	289.444	28.092	9.026	52.809	-288.484	28.166	9.167	52.888	-289.437
0.0	0.0	-2.167	29.833	0.002	0.028	-2.154	29.768	0.0	0.0	-2.166	29.388
0.0	0.0	-29.833	2.167	-0.013	0.039	-29.711	-1.976	0.0	0.0	-29.833	-2.167

(b) Approximations to $x(h, \varepsilon), h = 0.1$

Approximation	Approximating vector			
$e^{\bar{B}h}\zeta$	-55.4	-26.5	-1.31	-5.98
$e^{\bar{B}_i h}\zeta$	-55.2	-26.6	-1.28	-5.97
$e^{\bar{B}_{ii}h}\zeta$	-55.4	-26.5	-1.31	-5.98
$Ex(0.002)$	1.09	-13.9	-0.701	-0.873
$Ex(0.001)$	1.546	-13.6	-0.761	-1.01

TABLE 6.3-1 : Case 3 (Contd.)

(c) Approximations to $x(h, \varepsilon)$ $h = 0.05$

Approximation	Approximating vector			
$e^{\bar{B}h}\zeta$	-22.4	-10.9	1.12	-1.55
$e^{\tilde{B}h}\zeta$	-22.3	-10.9	1.12	1.54
$e^{\bar{B}_{ii}h}\zeta$	-22.4	-10.9	1.12	-1.55
$Ex(0.001)$	-14.2	-9.82	1.12	-1.54
$Ex(0.0005)$	-14.3	-9.77	0.965	-0.814

(d) Approximations to $x(h, \varepsilon)$, $h = 0.025$

Approximation	Approximating vector			
$e^{\bar{B}h}\zeta$	-9.46	-4.46	1.38	-0.0363
$Ex(0.00025)$	-8.46	-4.35	1.34	0.0557

TABLE 6.3-1: Case 4. $A = A(2, \tfrac{1}{4})$

(a) Approximations to $\bar{B}$

		$\delta_i = 0.01$	$T_i = 166.04$		$\delta_{ii} = 0.0001$	$T_{ii} = 150.8$

$\bar{B}(2,\tfrac{1}{4})$				$\bar{B}_i(2,\tfrac{1}{4})$				$\bar{B}_{ii}(2,\tfrac{1}{4})$			
6.3	−18.83	−133.02	−20.91	6.30	−19.06	−134.21	−20.18	6.30	−18.83	−133.02	−20.91
18.83	6.3	16.43	−131.42	18.92	6.20	17.35	−132.31	18.83	6.30	16.43	−131.42
0.0	0.0	0.70	20.83	−0.05	0.02	0.45	21.03	0.0	0.0	0.70	20.83
0.0	0.0	−20.83	0.70	−0.01	−0.05	−20.95	1.04	0.0	0.0	−20.83	0.70

(b) Approximations to $x(h, \varepsilon)$, $h = 0.1$

Approximation	Approximating vector			
$e^{\bar{B}h}\xi$	−24.6	−11.5	1.13	−3.33
$e^{\bar{B}_i h}\xi$	−24.8	−11.5	1.12	−3.32
$e^{\bar{B}_{ii} h}\xi$	−24.6	−11.5	1.13	−3.33
$Ex(0.0006)$	−10.8	−8.33	0.41	−1.46
$Ex(0.0007)$	−10.9	−8.35	0.41	−1.46

TABLE 6.3-1: Case 4 (*Contd.*)

(c) Approximations to $x(h, \varepsilon)$, $h = 0.05$

Approximation	Approximating vector		
$e^{\bar{B}h}\zeta$	-9.57	-4.37	1.57	-0.585
$e^{\bar{B}_* h}\zeta$	-9.63	-4.37	1.57	-0.579
$e^{\bar{B}_{1:h}}\zeta$	-9.57	-4.37	1.57	-0.585
$Ex(0.0003)$	-7.97	-3.95	1.42	-0.380
$Ex(0.00035)$	-7.94	-4.02	1.42	-0.37

(d) Approximations to $x(h, \varepsilon)$, $h = 0.025$

Approximation →	Approximation vector		
$e^{\bar{B}h}\zeta$	-3.72	-1.47	1.41	0.352
$Ex(0.00018)$	-3.53	-1.43	1.39	0.377

The average $\bar{B}$ can be determined in closed form, which we denote by $\bar{B}(1,w)$ and $\bar{B}(2,w)$ respectively. These are compared with $\bar{B}s$ obtained by methods (i) and (ii); e.g., $\bar{B}(1,w)$ is compared with $\bar{B}_i(1,w)$ and $\bar{B}_{ii}(1,w)$ in Table 6.3-1.

The averaging method gives $e^{\bar{B}h}\zeta$ as an approximation to $x(h,\varepsilon)$. Table 6.3-1 compares these approximations for the three ways of determining $\bar{B}$. The extrapolation formula (6.3.7) gives an approximation to $x(h,\varepsilon)$. Results of two utilizations of this formula are presented in Table 6.3-1 as well. These are denoted by $Ex(\varepsilon_i')$ and $Ex(\varepsilon_{ii}')$, respectively, where $\varepsilon_i' = h/T_i$ and $\varepsilon_{ii}' = h/T_{ii}$. We observe that no value of ε is prescribed for these computations. This demonstrates the effectiveness of singular perturbation methods in supplying approximations which are uniformly valid for all ε smaller than some prescribed value ε_0, say. The latter in turn depends only on the accuracy desired of the leading term of the expansion as an approximation to the full solution (for the particular differential equation at hand). (Compare the comments following (5.3.7).)

6.3.5. Discussion

Existing stiff differential equation solving routines, such as Gear's (see Hindmarsh, 1974), can degrade markedly when applied to problems having highly oscillatory solutions since computations must continually be made using very small increments. On the other hand, methods like those presented here can require extensive *a priori* preparation of the system to be solved.

Numerical implementation of the averaging procedure requires the determination of T_i (Section 6.3.2.i) which is used in direct approximation of $\bar{f}$ or determination of T_{ii} (Section 6.3.2.ii) as an approximate translation number for the integral $V(T,x)$.

The ratio h/ε is a measure of the system's stiffness, while $1/(\varepsilon h)$ measures the work involved in direct computation of solutions. On the other hand, $T_i/\Delta T$ (ΔT is the increment used in the averaging quadrature) measures the work needed to calculate $\bar{f}$, and $T_{ii}/\Delta T$ is a measure of the work involved in calculating the approximate translation number. While the method based on approximation of $\bar{f}$ in Section (6.3.2(i)) is reliable, it is costly even for a periodic function f. The work involved in the computation of translation numbers varies from minimal (e.g., for a periodic function) to an amount which offers no improvement when resonances occur in the integral V.

Finally, note that formula (6.3.7) shows the error arising in the extrapolation procedure decreases as a power of h; e.g., replacing h by $h/2$ implies the error changes from h^2/ε to $h^2/(4\varepsilon)$. Consequences of this fact are illustrated in Table 6.3-1 where computations are carried out for the two or three values of h. For example, from Case 3 of that table we may calculate the following:

$$\| e^{0.1B}\xi - Ex(0.001) \| = 59.5,$$
$$\| e^{0.05B}\xi - Ex(0.0005) \| = 8.1,$$
$$\| e^{0.025B}\xi - Ex(0.00025) \| = 1.1.$$

Here $\| \cdot \|$ denotes the Euclidean norm in $\mathbf{R}^4$.

6.4. A METHOD OF AVERAGING

6.4.1. *Motivation: Stable Functionals*

Consider the following model problem

$$\ddot{x} + \lambda^2 x = \lambda^2 \sin t, \tag{6.4.1}$$

and the following family of solutions

$$x(t) = a \sin \lambda t + \frac{\sin t}{1 - 1/\lambda^2}. \tag{6.4.2}$$

For λ large, this solution family consists of a high frequency *carrier wave*, $a \sin \lambda t$, modulated by a *slow wave*, $(\sin t)/(1 - 1/\lambda^2)$. As we have observed, the specification of the value at a point of such a function is an ill-conditioned problem.

We have seen that the linear multistep class of methods is highly desirable for numerical analysis since these methods are easy to use and easy to analyze. However these methods consist of a linear combination of *unstable functionals* of the solution of (6.4.1), namely values and values of derivatives at points. In this section, we show how to replace these unstable functionals by stable ones, thereby producing a class of linear multistep methods suitable for the stiff problem. We suppose that the stable functionals provide information about the solution being sought, and (subject to a process like mesh refinement) that the stable functionals furnish as adequate a description of the solution as needed.

We do not characterize the classes of functionals which are stable in an abstract way. Instead we select two special functionals, an averaging functional and an appropriate evaluation functional, which are stable in the sense discussed. We construct the numerical methods out of these two functionals.

6.4.2. *The Problem Treated*

We develop the method in the context of the problem,

$$\ddot{x} + \lambda^2 x = f(x, t), \qquad t \in [0, T],$$
$$x(0) = x_0, \tag{6.4.3}$$

where x and f are scalars.

The solution of this problem will be required to exist on the larger interval $\mathscr{I} = [-\tau, T]$, where the positive quantity τ will be specified in (6.4.9). Thus, we assume that $f(x, t)$ is continuous in $t, t \in \mathscr{I}$ and Lipschitz continuous in x for all such t, with Lipschitz constant L. Then $f(x, t)$ is uniformly bounded for $t \in \mathscr{I}$ and x restricted to any compact set including, in particular, the set of values taken on by the solutions $x(t)$ for $t \in \mathscr{I}$. At first we restrict our attention to the linear problem in which $f(x, t) = f(t)$. Then in Section 6.4.9, we make some comments about the nonlinear case and the case of second order systems.

6.4.3. *Choice of Functionals*

Let r, s and N be positive integers, let $h = T/N$ and let $t_i = ih, i = 0, \pm 1, \ldots$ be the points of a mesh. Let $z(t)$ be a functional of x which can be calculated at each mesh point. Then we seek to determine $y_n = y(t_n)$, in terms of $y_{n-i}, i = 1, \ldots, r$ and $z_{n-1} = z(t_{n-i}), i = 0, 1, \ldots, s$ by means of the linear multistep formula

$$\sum_{i=0}^{r} a_i y_{n-1} + \sum_{i=0}^{s} b_i z_{n-i} = 0, \qquad n = 0, 1, \ldots, N. \tag{6.4.4}$$

The initial values $y_i, i = -1, \ldots, -r$ are assumed to be furnished by some independent means, i.e., by a starting procedure (compare (1.2.5)f.).

In the case (6.4.3) of interest and λ large, we choose $y(t)$ to be

$$y(t) = \int_{-\infty}^{\infty} k(t - s) x(s) ds, \tag{6.4.5}$$

where

$$k(z) = \frac{1}{\Delta} \begin{cases} 1, & -\Delta < z < 0, \\ 0, & \text{otherwise.} \end{cases} \tag{6.4.6}$$

Thus $y(t)$ represents the average of $x(t)$ over the interval $[t - \Delta, t]$.

The functional $z(t)$ is chosen to be $[d^2/dt^2 + \lambda^2]x(t)$, i.e., $f(t)$, which we suppose can be stably calculated at each mesh point. Thus with a change in normalization, (6.4.4) may be written as

$$y_n = \sum_{i=1}^{r} c_i y_{n-i} + h^2 \sum_{i=0}^{s} d_i f_{n-i}. \tag{6.4.7}$$

6.4.4. Representers

We introduce the *reproducing kernel space*, $\mathscr{H} \equiv \mathscr{H}_m$ which is the Sobolev space $W_m^2[-\infty, \infty]$ with the inner product

$$\langle f, g \rangle = \sum_{j=0}^{m} \binom{m}{j} (f^{(j)}, g^{(j)}), \tag{6.4.8}$$

where

$$(f, g) = \int_{-\infty}^{\infty} f(t)g^*(t)dt.$$

An asterisk is used to denote the complex conjugate throughout. Since we are interested in solutions of (6.4.3) on the interval

$$\mathscr{I} = [-r\Delta, T], \tag{6.4.9}$$

we may identify both a solution of (6.4.3) and $f(t)$ appearing in (6.4.3) with the unique functions of minimal norm in $\mathscr{H}$ with which they agree on $\mathscr{I}$, respectively. Of course on $\mathscr{I}$, f is required to have $m-1$ absolutely continuous derivatives and an mth derivative (almost everywhere) which is square integrable.

We use a carat to denote the Fourier transform, viz.

$$f(t) = \frac{1}{\sqrt{2\pi}} \int_{-\infty}^{\infty} e^{i\omega t} \hat{f}(\omega)d\omega, \qquad \hat{f}(\omega) = \frac{1}{\sqrt{2\pi}} \int_{-\infty}^{\infty} e^{-i\omega t} f(t)dt. \tag{6.4.10}$$

Then the inner product in $\mathscr{H}$ may be written as

$$\langle f, g \rangle = \frac{1}{\sqrt{2\pi}} \int_{-\infty}^{\infty} \hat{f}(\omega)\hat{g}^*(\omega) |P_m(\omega)|^2 d\omega, \tag{6.4.11}$$

where

$$P_m(\omega) = (1 - i\omega)^m. \tag{6.4.12}$$

The *reproducing kernel* in $\mathcal{H}$ is (see Michelli and Miranker, 1973)

$$R_t \equiv R_t^m(s) = \frac{1}{\sqrt{2\pi}} \int_{-\infty}^{\infty} \frac{e^{i(s-t)\omega}}{|P_m(\omega)|^2} d\omega. \tag{6.4.13}$$

A second Hilbert space, $\hat{\mathcal{H}}$ is introduced as follows.

$$\hat{\mathcal{H}} \equiv \hat{\mathcal{H}}_m = \{\hat{f} \,|\, \hat{f} P_m \in L^2[-\infty, \infty]\}. \tag{6.4.14}$$

The inner product in $\hat{\mathcal{H}}$ is

$$\langle \hat{f}, \hat{g} \rangle = \frac{1}{\sqrt{2\pi}} \int_{-\infty}^{\infty} \hat{f} \hat{g}^* |P_m|^2 d\omega. \tag{6.4.15}$$

(6.4.11) defines an *isometric isomorphism* between $\mathcal{H}$ and $\hat{\mathcal{H}}$. We use the symbol $\sim$ to denote this isomorphism. Then from (6.4.11), we see that the isomorphism between R_t and its image in $\hat{\mathcal{H}}$ is expressed as follows.

$$R_t \sim \frac{e^{-i\omega t}}{|P_m(\omega)|^2}. \tag{6.4.16}$$

Then for the *representer* η_t of $(d^2/dt^2) + \lambda^2$, we have

$$\eta_t \equiv R_t'' + \lambda^2 R_t \sim (-\omega^2 + \lambda^2) \frac{e^{-i\omega t}}{|P_m(\omega)|^2}. \tag{6.4.17}$$

For the representer k_t of $y(t)$ given by (6.4.5) and (6.4.6), we have

$$\begin{aligned}
k_t \equiv k_t(s) &= \frac{1}{\Delta} \int_{t-\Delta}^{t} R_u(s) du \\
&\sim \frac{1}{\Delta} \int_{t-\Delta}^{t} \frac{e^{-i\omega u}}{|P_m(\omega)|^2} du \\
&= \frac{1}{|P_m(\omega)|^2} e^{-i\omega t} \left[\frac{1 - e^{-i\omega\Delta}}{-i\omega\Delta} \right] \\
&= \frac{e^{-i\omega t}}{|P_m(\omega)|^2} \sqrt{2\pi} \hat{k}(\omega),
\end{aligned} \tag{6.4.18}$$

where $\hat{k}(\omega)$ is the Fourier transform of the function $k(z)$ given in (6.4.6).

With these representers, the formula (6.4.7) leads us to introduce the following linear functional g_n.

$$g_n \equiv g_n[x] \equiv \left\langle k_{t_n} - \sum_{i=1}^{r} c_i k_{t_{n-i}} - h^2 \sum_{i=0}^{s} d_i \eta_{t_{n-i}}, x \right\rangle. \tag{6.4.19}$$

g_n will be zero if x is the numerical solution. In general g_n is not zero and is the analogue of the local truncation error for classical linear multistep schemes.

6.4.5. Local Error and Generalized Moment Conditions

g_n is characterized in the following definition.

DEFINITION 6.4.1. Using (6.4.19) as a definition, we call the linear functional, g_n appearing there the *local truncation error (functional)* of the method (6.4.7). The *(generalized) local truncation error* is $\|g_n\|^2$, where

$$\|x\|^2 = \langle x, x \rangle \quad \text{and} \quad \|x\|_\wedge^2 = \langle x, x \rangle_\wedge. \tag{6.4.20}$$

To estimate the local truncation error we write

$$\|g_n\|^2 \leq \left\| k_{t_n} - \sum_{j=1}^r c_j k_{t_{n-j}} - h^2 \sum_{j=0}^s d_j \eta_{t_{n-j}} \right\|_\wedge^2, \tag{6.4.21}$$

We will drop the subscript $\wedge$, since no confusion should result. Now using (6.4.15), (6.4.18) and (6.4.19), we find for the right member of (6.4.21) that

$$\left\| k_{t_n} - \sum_{j=1}^r c_j k_{t_{n-j}} - h^2 \sum_{j=0}^s d_j \eta_{t_{n-j}} \right\|^2 = \frac{1}{\sqrt{2\pi}} \int_{-\infty}^\infty |t(\omega)|^2 \frac{d\omega}{|P_m(\omega)|^2}, \tag{6.4.22}$$

where

$$t(\omega) = \frac{1}{\sqrt{2\pi}} \hat{k}(\omega) \sum_{j=0}^r s_j e^{ij\omega h} - h^2(\lambda^2 - \omega^2) \sum_{j=1}^s d_j e^{ij\omega h}. \tag{6.4.23}$$

Here

$$s_0 = 1 \quad \text{and} \quad s_j = -c_j, \quad j = 1, \ldots, r. \tag{6.4.24}$$

Expanding $t(\omega)$ formally in a Taylor series with remainder gives

$$t(\omega) = \sum_{l=1}^{p-1} (ih\omega)^l m_l + R_p, \tag{6.4.25}$$

where from (6.4.23) and (6.4.25), we obtain

$$m_l = \frac{1}{(l+1)!} \sum_{k=1}^{l+1} \binom{l+1}{k} L^{k-1} \sum_{j=0}^r j^{1+l-k} s_j$$
$$- \frac{h^2 \lambda^2}{l!} \sum_{j=0}^s j^l d_j - \frac{1}{(l-2)!} \sum_{j=0}^s j^{l-2} d_j \tag{6.4.26}$$

and

$$R_p = \frac{(ih\omega)^p}{p!} \left[-\frac{1}{L(p+1)} \sum_{j=0}^{r} s_j (j^{p+1} e^{ijh\omega_{j,1}} - (j+L)^{p+1} \right.$$

$$\times e^{ijh(1+L)\omega_{j,2}}) - h^2\lambda^2 \sum_{j=0}^{s} j^p d_j e^{ijh\omega_{j,3}} - p(p-1)$$

$$\left. \times \sum_{j=0}^{s} j^{p-2} d_j e^{ijh\omega_{j,4}} \right]. \tag{6.4.27}$$

In (6.4.26) and (6.4.27), we use the abbreviation

$$L = \Delta/h. \tag{6.4.28}$$

That is, in terms of the functional k of (6.4.5) and (6.4.6), the interval Δ, over which the average is taken, is a multiple L, of the mesh increment h. In (6.4.27) the quantities $\omega_{j,1}$ and $\omega_{j,2}$, $j = 0, \ldots, r$ and $\omega_{j,3}$ and $\omega_{j,4}$, $j = 0, \ldots, s$ are of values of ω which arise from the calculation of the remainder in Taylor's theorem.

The quantities m_l are characterized in the following definition.

DEFINITION 6.4.2. We call the m_l, $l = 0, 1, \ldots$, the *generalized moments* (of the coefficients). Analogously $m_l = 0$, $l = 0, 1, \ldots$, will be called the *generalized moment conditions*.

We make the following observation concerning the determination of the coefficients of the numerical method from the generalized moment conditions.

REMARK 6.4.3. View the equations $m_l = 0$, $l = 0, \ldots, r-1$ as r equations for the r unknowns s_j, $j = 1, \ldots, r$. The lth row of the resulting coefficient matrix which has as its jth term

$$\frac{1}{(l+1)!} \sum_{k=1}^{l} \binom{l+1}{r} L^{k-1} m^{1+l-k},$$

is a linear combination of the first l rows of the Vandermonde matrix. Thus the system of r equations has a solution in this case. Indeed by choosing the d_j, $j = 0, \ldots, s$ to be proportional to λ^{-2}, we obtain a solution for the s_j, $j = 1, \ldots, r$ which is $O(1) + O(\lambda^{-2})$.

From the form of $t(\omega)$ given in (6.4.23), we may make the following remark, the assertion in which follows from a familiar argument which proceeds by breaking up the range of integration in (6.4.22) appropriately.

REMARK 6.4.4. If p is chosen less than m, and the coefficients $s_j, j = 1, \ldots, r$ and $d_j, j = 0, \ldots, s$ are chosen as solutions of the generalized moment conditions $m_l = 0$, $l = 0, 1, \ldots, p$, we may obtain an estimate of the local truncation error of the following form.

$$\| g_n \| \equiv \max_{\substack{x \in \mathscr{H} \\ \|x\| \leq 1}} |g_n| \leq O(h^{p+1}), \quad p < m. \tag{6.4.29}$$

We collect these remarks into the following theorem.

THEOREM 6.4.5. *There exists a choice of coefficients* $s_j, j = 1, \ldots, r$ *and* $d_j, j = 0, \ldots, s$, *such that the local truncation error has a bound of the form* (6.4.29). *Moreover, this bound is uniform in* λ *for* $|\lambda| \geq \lambda_0$ *for any fixed positive* λ_0.

6.4.6. Stability and Global Error Analysis

y_n, $n = 0, 1, \ldots$ denotes the values obtained by the multistep formula (6.4.7) from the initial values $y_n, n = -r, \ldots, -1$. Let $Y_n, n = -r$, $-r + 1, \ldots$ denote the exact values of these functionals. Let

$$e_n = y_n - Y_n, \qquad n = -r, -r + 1, \ldots$$

denote the global error. For convenience, assume that the initial functionals vanish, i.e., $e_n = 0, n = -r, -r + 1, \ldots, -1$.

Subtract the following identity

$$Y_n = \sum_{j=1}^{r} c_j Y_{n-j} + h^2 \sum_{j=0}^{s} d_j f_{n-j} + Y_n - \sum_{j=1}^{r} c_j Y_{n-j} - h^2 \sum_{j=0}^{s} d_j f_{n-j},$$

from (6.4.7). We get

$$e_n = \sum_{j=1}^{r} c_j e_{n-j} + g_n. \tag{6.4.30}$$

Here

$$g_n = -y_n + \sum_{j=1}^{r} c_j Y_{n-j} + h^2 \sum_{j=0}^{s} d_j f_{n-j},$$

is the value of the linear functional g_n of (6.4.19) applied to x, the latter being the exact solution of the initial value problem (6.4.3). To solve (6.4.30) for e_n, we use the polynomial $S(z)$ (compare (4.2.18)):

$$S(z) = \sum_{j=0}^{r} s_j z^{r-j}.$$

Since $s_0 = 1$, $[z^r S(z^{-1})]^{-1}$ is an analytic function of z in a neighborhood of $z = 0$. Then let its power series be given by

$$[z^r S(z^{-1})]^{-1} = \sum_{j=0}^{\infty} \sigma_j z^j.$$

Now multiply (6.4.30) by σ_{N-n} and sum the result over n from $n = r$ to N. The result is the solution of (6.4.30) (compare (4.2.19)–(4.2.22)):

$$e_N = \sum_{n=r}^{N} \sigma_{N-n} g_n. \tag{6.4.31}$$

Stability of these methods is characterized in the following definition.

DEFINITION 6.4.6.   If the sequence $\{\sigma_j, j = 0, 1, \dots\}$ is bounded, then the method is said to be stable.

We recall the following definition.

DEFINITION 6.4.7. $S(z)$ is said to obey the root condition if all of its roots lie in the closed unit disc while those of its roots which lie on the boundary of that disc are simple.

With this definition we may state the following lemma which characterizes the stability of the method.

LEMMA 6.4.8. *If the polynomial $S(z)$ obeys the root condition, then the sequence $\{\sigma_j, j = 0, 1, \dots\}$ is bounded, i.e., the method is stable. (See Lemma 4.2.2).*

If Lemma 6.4.8 is applicable, (6.4.31) gives

$$|e_N| \leq \text{const} \times N \max_{r \leq n \leq N} \|g_n\| \|x\|,$$

where x is the exact solution of (6.4.3).

Combining this with (6.4.28) gives the following convergence theorem for the method.

THEOREM 6.4.9. *If the choice of coefficients characterized in Lemma 6.4.8 gives rise to a stable method, then for the method (6.4.7) with those coefficients,*

$$\|e_N\| \leq O(h^p), \qquad p < m, \tag{6.4.32}$$

uniformly in λ for $|\lambda| \geq \lambda_0$ for any fixed positive λ_0.

6.4.7. *Examples*

We now consider some examples of methods of the type (6.4.7) in which the coefficients are determined by the generalized moment conditions.
From (6.4.26) we have for $l = 0$, 1 and 2, respectively,

$$(0) \quad m_0 \equiv \sum_{j=0}^{r} s_j - h^2 \lambda^2 \sum_{j=0}^{s} d_j,$$

$$(1) \quad m_1 \equiv \sum_{j=0}^{r} j s_j + \frac{L}{2} \sum_{j=0}^{r} s_j - h^2 \lambda^2 \sum_{j=0}^{s} j d_j, \qquad (6.4.33)$$

$$(2) \quad m_2 \equiv \frac{1}{2} \sum_{j=0}^{r} j^2 s_j + \frac{L}{2} \sum_{j=0}^{r} j s_j + \frac{L^2}{6} \sum_{j=0}^{r} s_j - \frac{h^2 \lambda^2}{2} \sum_{j=0}^{s} j^2 d_j - \sum_{j=0}^{s} d_j.$$

Consider the following case where the first two generalized moment conditions are satisfied.

(A) $m_0 = m_1 = 0$.
  For $r = s = 1$, we get

$$c_1 = 1 - \frac{2}{L} + \frac{2}{L} h^2 \lambda^2 d_0,$$

$$d_1 = \frac{2}{h^2 \lambda^2 L} - \left(\frac{2}{L} + 1 \right) d_0. \qquad (6.4.34)$$

In the special case $d_0 = 0$, (6.4.34) becomes

$$I \begin{cases} c_1 = 1 - \dfrac{2}{L}, \\[2mm] d_1 = \dfrac{2}{h^2 \lambda^2 L}. \end{cases}$$

These coefficients (i.e., c_1) obey the root condition if and only if $L \geq 1$.
In the special case $d_0 = d_1$, (6.4.34) becomes

$$II \begin{cases} c_1 = 1 - \dfrac{2}{L+1}, \\[2mm] d_0 = d_1 = \dfrac{1}{h^2 \lambda^2} \dfrac{1}{L+1}. \end{cases} \qquad (6.4.35)$$

Under the restriction $L \geq 0$, the root condition is equivalent to $L \geq 0$ for the coefficients (6.4.35).

For $r = s = 2$,

$$c_1 = 1 - \frac{2}{L} - \left(1 + \frac{2}{L}\right)c_2 + \frac{2}{L}\lambda^2 h^2 (d_0 - d_1),$$

$$\text{(6.4,36)}$$

$$d_1 = \frac{2}{\lambda^2 h^2 L}(1 + c_2) - \left(1 + \frac{2}{L}\right)d_0 - \left(1 - \frac{2}{L}\right)d_2.$$

In the special case $d_0 = 0, c_1 = c_2, d_1 = d_2$, (6.4.36) becomes

$$III \begin{cases} c_1 = c_2 = \dfrac{L-3}{2L} \\[2mm] d_1 = d_2 = \dfrac{3}{2\lambda^2 h^2 L}. \end{cases}$$

In this case,

$$S(z) = z^2 - \frac{L-3}{2L}z - \frac{L-3}{2L},$$

and this polynomial $S(z)$ obeys the root condition for a set of values of L which includes all $L \geq 1$.

In the special case $c_1 = c_2, d_1 = d_2 = 0$, (6.4.36) becomes

$$IV \begin{cases} c_1 = c_2 = \dfrac{1}{2}\dfrac{L}{3+L}, \\[2mm] d_0 = \dfrac{1}{\lambda^2 h^2}\dfrac{3}{3+L}. \end{cases}$$

Here

$$S(z) = z^2 - \frac{1}{2}\frac{L}{3+L}z - \frac{1}{2}\frac{L}{3+L}.$$

This polynomial obeys the root condition for a set of values of L which includes all $L > 0$.

In the special case $c_1 = c_2, d_0 = d_1 = d_2$ (6.4.36) becomes

$$V \begin{cases} c_1 = c_2 = \dfrac{1}{2}\dfrac{L-1}{L+1}, \\[2mm] d_0 = d_1 = d_2 = \dfrac{2}{3\lambda^2 h^2}\dfrac{1}{1+L}. \end{cases}$$

In this case, the root condition is obeyed for $L > 0$.

Now we consider a case corresponding to three generalized moment conditions.

(B) $m_0 = m_1 = m_2 = 0$.
For $r = s = 1$, we get

$$VI \begin{cases} c_1 = 1 - L\left(\dfrac{L^2}{3} + \dfrac{L}{2} - \dfrac{2}{h^2\lambda^2}\right)^{-1}, \\[3mm] d_0 = \dfrac{1}{\lambda^2 h^2}\left[1 - L + L^2\left(\dfrac{2}{3}L^2 + L - \dfrac{4}{h^2\lambda^2}\right)^{-1}\right], \\[3mm] d_1 = \dfrac{1}{\lambda^2 h^2}\left[-1 + L + (2L - L^2)\left(\dfrac{2}{3}L^2 + L - \dfrac{4}{h^2\lambda^2}\right)^{-1}\right]. \end{cases} \qquad (6.4.50)$$

Notice that the root condition is obeyed for L large and positive but is violated for $h\lambda$ small compared to L.

We make the following observations concerning the explicit dependence of the coefficients of the numerical methods on the coefficients of the differential equation, in particular on λ^2.

REMARK 6.4.10. In all of these examples as in the general case, we see that the coefficients obtained as solutions of the moment conditions depend on λ^2. At first sight this seems to be more restrictive than the case of the classical linear multistep formulas where the coefficients of the formula do not depend on the coefficients of the differential equation. In fact there is no such distinction. In the classical case, the coefficients of the differential equation enter into the method when it is used to approximate the differential equation, e.g., when $\dot{y}_{n-i}$ is replaced by $f(y_{n-i}, t_{n-i})$. It is essential after all that the numerical method at some point be dependent on the equation to be solved. In the present development, this dependence occurs at the outset in the determination of coefficients and in the error analysis. In the classical case it enters in the error analysis and in the use of the methods.

A generalization of the methods described here which also utilize concepts of the two-time approach of Section 6.1 can be found in Miranker and van Veldhuizen, 1978.

6.4.8. Computational Experiments

We now apply the six sets of methods labeled I, II, ... , VI in Section 6.4.7, to the model problem:

$$\ddot{x} + \lambda^2 x = \lambda^2 \sin t,$$

$$x(0) = 0, \quad x'(0) = \frac{\lambda}{2} + \frac{1}{1 - 1/\lambda^2}.$$

Computations are made over the interval $[0, T] = [0, \pi]$. In Table 6.4-1, we display the $h^{1/2} \times l^2$-norm of the global error:

$$\|e\|_{l^2} \equiv \left[h \sum_{n=0}^{[\pi/h]} e_n^2 \right]^{1/2},$$

for a set of various combinations of $h = 0.1, 0.01, \lambda = 10, 10^3, 10^5$ and

TABLE 6.4-1

Method	$\lambda \backslash L$	1	2	3	1	2	3
I	10	0.273	0.108	0.112	0.133	0.126	0.126
	10^3	0.113	0.00217	0.0611	0.0283	0.00683	0.0083
	10^5	0.112	0.00209	0.0611	0.0111	0.000106	0.00627
II	10	0.122	0.133	0.155	0.126	0.127	0.128
	10^3	0.00125	0.0622	0.177	0.0241	0.00926	0.0136
	10^5	0.00104	0.0621	0.177	0.000118	0.00627	0.0125
III	10	0.242	0.111	0.0872	0.136	0.126	0.126
	10^3	0.0032	0.00422	0.00317	0.0294	0.00684	0.00546
	10^5	0.0034	0.00419	0.00313	0.00023	0.00112	0.89E-6
IV	10	0.123	0.111	0.0938	0.126	0.126	0.126
	10^3	0.00627	0.0144	0.0244	0.0241	0.00684	0.00546
	10^5	0.00623	0.0144	0.0244	0.000133	0.000179	0.000264
V	10	0.144	0.152	0.156	0.127	0.127	0.128
	10^3	0.0657	0.094	0.119	0.0249	0.0116	0.0136
	10^5	0.0657	0.0939	0.119	0.0063	0.00942	0.0125
VI	10	0.758E4	0.66E11	0.124	0.195E1	0.471E1	0.11E2
	10^3	0.0447	0.0639	0.244	0.0246	0.00901	0.0253
	10^5	0.0447	0.0639	0.244	0.00421	0.00629	0.0251
h		0.1			0.01		

$$h^{1/2} \|e\|_{l^2}$$

$L = 1, 2, 3$ and for each of the six methods cited. Here $[\pi/h]$ denotes the integer part of π/h.

To illustrate both the favorable and unfavorable effects, Table 6.4-1 contains cases for which the methods are designed to operate well along with cases with poor or nonsensical results.

For example although the cases corresponding to $\lambda = 10$ give fair results, these cases are not stiff, and we should not expect good results. When h is decreased, improvement should occur but only for the stiff cases. The cases $\lambda = 10^3$ and $h = 0.01$ are not stiff, and improvement with decreasing h does not always occur in these cases. Method VI is used in some unstable cases. Examining (6.4.27.) we see that R_p is proportional to L^p (see (6.4.28)). Thus in some cases as L increases, we see an improvement due to improving the averaging (i.e., increasing Δ), but ultimately a degradation due to the L dependence of R_p. The stiff cases for moderate L give extremely good results as we expect.

6.4.9. *The Nonlinear Case and the Case of Systems*

In Miranker and Wahba, 1976, a discussion of the extension of the results described in Sections 6.4.1–6.4.8 to the nonlinear case and to the case of systems is given. We give a survey of that discussion.

In the nonlinear case, f_{n-i} in the multistep formula (6.4.7) is replaced by $f(y_{n-i}, t_{n-i})$, since $f_{n-i} = f(x_{n-i}, t_{n-i})$ cannot be computed as we proceed along with the mesh. This results in a degradation of the error estimate (6.4.32) to the following one.

$$\| e_N \| \leq \text{const} \times [h^p + L\varepsilon_1 v_m]. \tag{6.4.37}$$

Here

$$\varepsilon_1 = \max_j |h^2 d_j L|,$$

where, as before, L is the Lipschitz constant of f, and

$$v_m = \frac{1}{2} \left[\int_{-\infty}^{\infty} \frac{|\omega|^2}{|P_m(\omega)|^2} \, d\omega \right]^{1/2}.$$

We make the following observation concerning this error estimate.

REMARK 6.4.11. The two terms in the estimate (6.4.37) are not comparable in orders of h. The first term, which corresponds to the local truncation error, is small for h small. The second term is the error by which a function may be approximated by its average. We may expect the latter to be small if λ is large. (6.4.27) may be viewed as the statement that modulo the error made in replacing a function by its average, the numerical method is globally h^p.

In the systems case, the differential equation (6.4.3) is replaced by the second order system

$$\ddot{x} + \Lambda^2 x = f(x, t).$$

Here x and f are q-vectors and Λ is a $q \times q$ matrix. The coefficients c_j (and s_j) and d_j of the numerical method are replaced by $q \times q$ matrices (denoted by the same symbols). Many such formal replacements of the scalar development follow. For example, the first two generalized moments become

$$m_0 = \left(\sum_{j=0}^{r} s_j - h^2 \Lambda^2 \sum_{j=0}^{s} d_j \right) \zeta_q,$$

$$(6.4.38)$$

$$m_1 = \left(\sum_{j=0}^{r} j s_j + \frac{L}{2} \sum_{j=0}^{r} s_j - \Lambda^2 \sum_{j=0}^{s} j d_j \right) \zeta_q$$

(compare (6.4.27)), where ζ_q is the q-vector all of whose components are unity.

The error analysis proceeds similarly (using some of the matricial arguments of Section 4.3, leading to an estimate of the global error which is similar to the one described in Theorem 6.4.9).

We conclude this summary of the systems case with the following two observations about these matricial methods.

REMARK 6.4.12. Referring to Remark 6.4.10 and noting the dependence of the coefficients of the numerical method on the coefficients of the differential equation, we see from (6.4.38) the way in which the dependence appears in terms of the matrix Λ^2, for the coefficients determined by the generalized moment conditions. It is important to take note that the coefficients depend on the matrix Λ^2 and not explicitly on eigenvalues of Λ^2. Thus, if we know that a system is stiff, with highly oscillatory components, we may use the methods described here without having to calculate the eigenvalues of Λ^2 which cause this stiffness.

REMARK 6.4.13. In the usual systems case for the numerical treatment of differential equations, the methods frequently used are the scalar methods with the scalar coefficients simply multiplied by the q-dimensional identity matrix. We conjecture that the methods developed here in the scalar case would work in the same way with the simple additional requirement of replacing λ or λ^{-1} by Λ or Λ^{-1}, respectively.

Chapter 7

Other Singularly Perturbed Problems

Summary

Thus far we have dealt with the numerical treatment of the initial value problem. There certainly are many other kinds of stiff and singularly perturbed problems, and we conclude this monograph with a discussion of two of them. These illustrate the wide variety of problems of this nature for which numerical methods have barely begun to be developed.

We begin in Section 7.1 with a study of recurrences containing a small parameter. Such recurrences arise in a number of applications several of which are illustrated. Then in Section 7.2 we turn to a model two point boundary value problem containing a small parameter. Moreover, we consider the cases in which turning points are present as well. Such problems are very well-known and arise, for example, by separation of variables in problems from mathematical physics.

7.1. SINGULARLY PERTURBED RECURRENCES

7.1.1. *Introduction and Motivation*

Numerical methods for differential equations which are neither stiff nor singularly perturbed proceed by replacement of the differential equation by a difference equation. In the stiff or singularly perturbed case, such a direct approach will result in a difference equation which is itself stiff or singularly perturbed and therefore of problematical efficacy. Suppose nevertheless, that we do proceed in this way for the singularly perturbed problem. The resulting difference equation will have a small parameter causing its solution to change rapidly. As with differential equations, these rapid changes are often superimposed upon a slowly changing aspect of the solution. As we have seen in a number of cases, it is often the slowly changing part of the solution which gives the important features.

164

Systems of difference equations or recurrences which contain small parameters arise not only through discretization of differential equations but also in fact in their own right in many applications of mathematics to physical, biological and engineering problems. Of course in the non-singularly perturbed case (the nonstiff case), the numerical solution of such recurrences is more or less straightforward. We already know that this is not true in the singularly perturbed case, wherein the numerical development might require extraction of the slowly developing part of the solution.

To illustrate this, consider the following system of recurrences.

$$x_{n+1} = (A + \varepsilon B)x_n, \qquad x_0 \quad \text{given}, \tag{7.1.1}$$

where $x \in R^p$ and A and B are $p \times p$ constant matrices. (7.1.1) has the solution

$$x_n = (A + \varepsilon B)^n x_0.$$

Now suppose that ε is a small parameter. For convenience, we will temporarily suppose that A and B commute. Then

$$x_n = A^n (I + \varepsilon A^{-1} B)^n x_0.$$

Here and hereafter I is the p-dimensional identity matrix. For large values of n, say $n = K[1/\varepsilon] + \rho$ (here $[1/\varepsilon]$ denotes the integer part of $1/\varepsilon$), so that $(I + \varepsilon A^{-1} B)^n = \exp[A^{-1} BK][I + O(\varepsilon)]$, and so the solution may be written as

$$x_n = A^n \exp[A^{-1} B\varepsilon n] x_0 (1 + n\varepsilon O(\varepsilon)), \tag{7.1.2}$$

which is a useful approximation when $n = O(1/\varepsilon)$. Thus, the solution can be expressed asymptotically for ε near 0, by the product of a factor which varies rapidly (i.e., A^n) with n and one that varies slowly (i.e., $\exp(A^{-1} B\varepsilon n)$) with n.

Separation of the fast and slowly varying parts can sometimes lead to an acceleration in the development of the recurrence. To illustrate this, let us consider an explicit example arising in *pattern recognition* where a recurrence relation appears in connection with a so-called *training algorithm*. The algorithm determines a hyperplane separating two finite point sets **A** and **B** whose respective *convex hulls* are disjoint.

Let $\mathbf{A} = \{a_i, \ i = 1, \ldots, q\}$ and $\mathbf{B} = \{b_j, j = 1, \ldots, r\}$ be finite point sets in R^p. The problem is to find a vector $v = (v_1, \ldots, v_p)$ and a scalar c such that

$$v \cdot b_j < c < v \cdot a_i, \quad i = 1, \ldots, q, \quad j = 1, \ldots, r. \tag{7.1.3}$$

Here we use a dot to denote the inner product of two finite dimensional vectors.

Then the equation $v \cdot x - c = 0$ determines a hyperplane which separates the sets $\mathbf{A}$ and $\mathbf{B}$.

We imbed the problem in R^{p+1} by introducing the following augmented vectors. $w = (v_1, \ldots, v_p, -c)$, $a_i^* = (a_{i1}, \ldots, a_{ip}, 1)$, $i = 1, \ldots, q$ and $b_j^* = (b_{j1}, \ldots, b_{jp}, 1)$, $j = 1, \ldots, r$. Let $\mathbf{A}^* = \{a_i^*, i = 1, \ldots, q\}$ and $\mathbf{B}^* = \{b_j^*, j = 1, \ldots, r\}$.

The training algorithm is described as follows.

Training algorithm

Let $T = \{x_1, x_2, \ldots\}$ by any sequence of vectors chosen from $\mathbf{A}^* \cup \mathbf{B}^*$. For fixed $\theta > 0$ and a given $w_0 \in R^{p+1}$, we define $w_1, w_2, \ldots$, as follows:

$$w_{n+1} = w_n + x_n S\left(\frac{w_n \cdot x_n}{\theta}; x_n\right), \qquad (7.1.4)$$

where

$$S\left(\frac{w \cdot x}{\theta}; x\right) = \begin{cases} \left.\begin{array}{l} 1, w \cdot x \leq \theta \\ 0, w \cdot x > \theta \end{array}\right\} \text{and } x \in \mathbf{A}^*, \\ \left.\begin{array}{l} -1, w \cdot x \geq -\theta \\ 0, w \cdot x > -\theta \end{array}\right\} \text{and } x \in \mathbf{B}^*. \end{cases} \qquad (7.1.5)$$

This algorithm is characterized by the following theorem (see Greenberg and Konheim, 1964).

THEOREM 7.1.1 *Given θ, w_0 and a training set T, let $\{w_n\}$ be determined by (7.1.6). Then the sequence $\{w_n\}$ converges. In fact, there is an integer N such that $w_N = w_{N+1} = \ldots$ If each element in $\mathbf{A}^* \cup \mathbf{B}^*$ occurs in T infinitely often, then the augmented vector $w_N \equiv (v_N, -c)$ supplies a hyperplane which separates $\mathbf{A}$ and $\mathbf{B}$.*

Through the change of variables: $w_n = \theta z(n)$ and $\varepsilon = \theta^{-1}$, the recurrence relation (7.1.4) for w_n is cast into the following perturbation form.

$$z(n+1) = z(n) + \varepsilon x_n S(z(n) \cdot x_n; x_n),$$

while in the definition (7.1.5) of S, w is changed to z and θ is replaced by unity.

As an example of the training algorithm consider the sets $\mathbf{A} = \{1\}$ and $\mathbf{B} = \{-1\}$ in R. Then $\mathbf{A}^* = \{x_1^* = (1, 1)\}$ and $\mathbf{B}^* = \{x_2^* = (1-1, 1)\}$.

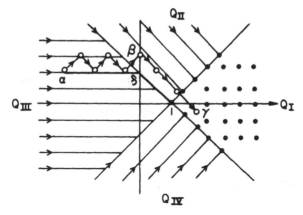

Fig. 7.1-1. Trajectories of the training algorithm $(\alpha\beta\gamma)$ and of its slowly varying part $(\alpha\delta 1)$.

The linear expressions $l_1(z) \equiv z \cdot x_1^* - 1$ and $l_2(z) \equiv z \cdot x_2^* + 1$ relevant to the definition of S, divide the z-plane into four quadrants Q_I, Q_{II}, Q_{III} and Q_{IV}, corresponding to the values of signum $(l_1, l_2) = (+, -), (+, +),$ $(-, +)$ and $(-, -)$, respectively. For $z \in Q_I, Q_{II}, Q_{III}$ and Q_{IV}, the pair $(S(z \cdot x_1^*; x_1^*), S(z \cdot x_2^*; x_2^*))$ takes on the values of $(0, 0), (0, -1), (1, -1)$ and $(1, 0)$, respectively. Choose T to be the periodic sequence $\{x_1^*, x_2^*, x_1^*, x_2^*, \ldots\}$.

In Figure 7.1-1, a sample trajectory for $z(n)$ which starts, at the point labeled α is plotted. This trajectory describes the sawtooth path $(\alpha\beta)$ and then the straight but segmented portion $(\beta\gamma)$. There are eight segments in all of this trajectory $(\alpha\beta\gamma)$, each one of length $\sqrt{2}\varepsilon$.

In Figure 7.1-1, we also plot a path $(\alpha\delta 1)$ consisting of two line segments. This latter path is the slowly changing part of the solution, i.e., of the trajectory $(\alpha\beta\gamma)$ of the training algorithm. Indeed as $\varepsilon \to 0$, we see that the corresponding sequence of training algorithm trajectories converges to $(\alpha\delta 1)$. Each such trajectory has the qualitative description of $(\alpha\beta\gamma)$ (i.e., a sawtooth portion followed by a straight segmented portion) except that the number of its segments tends to infinity.

This example leads us to guess that the limiting path $(\alpha\delta 1)$ is simply the trajectory which moves in the direction which the training algorithm trajectory itself takes, on the average. Indeed we will see that this is so. That is, the set of trajectories defined by the differential equation

$$dz_0/ds = \overline{xS}(z_0),\tag{7.1.6}$$

where the average

$$\overline{xS}(z_0) = \lim_{n \to \infty} \frac{1}{n} \sum_{k=0}^{n-1} x_k S(z_0 \cdot x_k; x_k),$$

determines the slowly changing part of the recurrence. (Compare (6.1.24) and (6.1.25)). In particular, we will see that $z(n) = z_0(\varepsilon n) + O(\varepsilon)$.

For this example, $\overline{xS}$ is conveniently computed since T is a periodic set. We have indeed that

$$2\overline{xS}(z) = \begin{cases} (0,0), & z \in Q_I, \\ (1,-1), & z \in Q_{II}, \\ (1,0), & z \in Q_{III}, \\ (1,1), & z \in Q_{IV}. \end{cases}$$

In Figure 7.1-1, we plot the direction fields of the differential equation (7.1.6). Notice that all points in Q_I are fixed points of the direction field, i.e., they are solutions of the set of inequalities (7.1.3).

7.1.2. The Two-time Formalism for Recurrences

The representation (7.1.2) of the development of the solution of the recurrence (7.1.1) on the time scales n and $s = \varepsilon n$ (*fast and slow times*, respectively) can be obtained by another method (compare Section 6.1) which is applicable to more general systems (systems with noncommuting A and B and nonlinear systems as well). We proceed by factoring out the dominant fast time behavior by setting

$$x_n = A^n u_n.$$

Thus u_n satisfies the following equation.

$$u_{n+1} = (I + \varepsilon A^{-1} B) u_n, \quad u_0 = x_0.$$

Since we expect the solution to depend on two time scales, n and $s = \varepsilon n$, we set

$$u_n = U(n, s, \varepsilon) = U^0(n, s) + \varepsilon U^1(n, s) + O(\varepsilon^2),$$

and attempt to determine the coefficients U^0 and U^1. To ensure that the expansion is meaningful as $\varepsilon \to 0$, we assume that the coefficients $U^0, U^1, \ldots,$ are bounded in n and s. Substituting this form into the equation for u_n gives

$$U(n+1, s + \varepsilon, \varepsilon) = (I + \varepsilon A^{-1} B) U(n, s, \varepsilon). \tag{7.1.7}$$

It follows that

$$U^0(n+1,s) = U^0(n,s), \tag{7.1.7}$$

$$U^1(n+1,s) = U^1(n,s) + A^{-1}BU^0(n,s) - \frac{\partial}{\partial s}U^0(n,s).$$

(7.1.7) shows that U^0 is independent of n, so we write $U^0 = U^0(s)$. The solution of (7.1.7) is

$$U^1(n,s) = U^1(0,s) + n[A^{-1}B\,U^0(s) - d\,U^0/ds].$$

Since U^1 is bounded, we must have,

$$\frac{dU^0}{ds} = A^{-1}BU^0, \quad U^0(0) = x_0.$$

Solving this equation and combining with the previous results, we are led to the following representation of x_n.

$$x_n = A^n \exp(A^{-1}B\varepsilon n)x_0 + O(\varepsilon).$$

Thus, the multi-time hypothesis gives a formal derivation of the earlier result (7.1.2). In the next section, this method is extended to more general systems where the slowly varying part of the solution will be derived by an averaging procedure.

7.1.3. The Averaging Procedure

The system of equations

$$x_{n+1} = Ax_n + \varepsilon f(x_n,\varepsilon), \quad x_0 \quad \text{given}, \tag{7.1.8}$$

will be studied under various conditions on A and f. Here $x, f \in R^p$, A is a $p \times p$ matrix and ε is a small positive parameter. Setting $p = q + r$, we now introduce the hypotheses H1, H2, H3 and H4 characterizing A and f.

H1: Suppose there exists an invertible matrix P such that

$$P^{-1}AP = \text{diag}(\mathcal{O}, \mathcal{S}),$$

where the $q \times q$ matrix $\mathcal{O}$ is *oscillatory*, i.e., has all characteristic roots on the unit disc $|\lambda| = 1$, in the complex λ-plane, and the $r \times r$ matrix $\mathcal{S}$ is *stable*, i.e., has all its characteristic roots inside the unit disc, $|\lambda| < 1$. (Contrast this with the notion of stable matrix introduced following (6.1.9).) The matrix $\mathcal{O}$ is assumed to be diagonalizable.

H2: Suppose the function f is a smooth function of its arguments.
We define new variables by means of the following relation.

$$x_n = P\begin{pmatrix} \mathcal{O}^n u_n \\ v_n \end{pmatrix}.$$

Here P and $\mathcal{O}$ are given in H1, and $u_n \in R^q, v_n \in R^r$. Then (7.1.8) becomes

$$\begin{cases} u_{n+1} = u_n + \varepsilon \mathcal{O}^{-n-1} g(\mathcal{O}^n u_n, v_n, \varepsilon), \\ v_{n+1} = \mathcal{S} v_n + \varepsilon h(\mathcal{O}^n u_n, v_n, \varepsilon), \end{cases}$$

where $f = P\begin{pmatrix} g \\ h \end{pmatrix}$, $g \in R^q$ and $h \in R^r$.

Next suppose that the equation here has a solution, viz.

H3: There is a smooth function $\phi(u, \varepsilon)$ such that

$$\phi = \mathcal{S}\phi + \varepsilon h(u, \phi, \varepsilon),$$

with $\phi = O(\varepsilon)$.

With this, we define $v_n = \phi(\mathcal{O}^n u_n, \varepsilon) + V_n$, and we obtain

$$\begin{cases} u_{n+1} = u_n + \varepsilon \mathcal{O}^{-n-1} g(\mathcal{O}^n u_n, \phi(\mathcal{O}^n u_n, \varepsilon) + V_n, \varepsilon), \\ v_{n+1} = \mathcal{S} v_n + \varepsilon[h(\mathcal{O}^n u_n, \phi(\mathcal{O}^n u_n, \varepsilon) + V_n, \varepsilon) \\ \quad - h(\mathcal{O}^n u_n, \phi(\mathcal{O}^n u_n, \varepsilon), \varepsilon)]. \end{cases}$$

To order ε, the u components will not change before the V components have equilibrated at $V_n = O(\varepsilon)$. Thus, in constructing the solution of the system, we consider the problem

$$u_{n+1} = u_n + \varepsilon \mathcal{O}^{-n-1} g(\mathcal{O}^n u_n, \phi(\mathcal{O}^n u_n, \varepsilon), \varepsilon), \quad u_0 = x_0.$$

The solution of this equation will be found in the form

$$u_n = U(n, s, \varepsilon) = U^0(n, s) + \varepsilon U^1(n, s) + O(\varepsilon^2),$$

where $s = \varepsilon n$. Therefore,

$$U(n + 1, s + \varepsilon, \varepsilon) = U(n, s, \varepsilon) + \varepsilon \mathcal{O}^{-n-1} g(\mathcal{O}^n U, \phi(\mathcal{O}^n U, \varepsilon), \varepsilon).$$

It follows that

$$U^0(n + 1, s) = U^0(n, s),$$

so $U^0(n, s) = U^0(s)$ is independent of n. Moreover,

$$U^1(n + 1, s) = U^1(n, s) + \mathcal{O}^{-n-1} g(\mathcal{O}^n U^0, 0, 0) - \frac{dU^0}{ds}.$$

Therefore,

$$U^1(n, s) = U^1(0, s) + \sum_{k=0}^{n-1} \mathcal{O}^{-k-1} g(\mathcal{O}^k U^0(s), 0, 0) - n \frac{dU^0}{ds}.$$

This and the boundedness of U^1 implies that

$$\frac{dU^0}{ds} = \lim_{n \to \infty} \frac{1}{n} \sum_{k=0}^{n-1} \mathcal{O}^{-k-1} g(\mathcal{O}^k U^0(s), 0, 0). \tag{7'1.9}$$

The next hypothesis ensures that equation (7.1.9) is meaningful.

H4: The average

$$\bar{g}(u) = \lim_{n \to \infty} \frac{1}{n} \sum_{k=0}^{n-1} \mathcal{O}^{-k-1} g(\mathcal{O}^k u, 0, 0),$$

exists and defines a smooth function of u. Moreover, suppose that the difference

$$\frac{1}{n} \sum_{k=0}^{n-1} \mathcal{O}^{-k-1} g(\mathcal{O}^k u, 0, 0) - \bar{g}(u),$$

remains bounded uniformly in n and u.

Note that if the characteristic roots of $\mathcal{O}$ are Mth roots of unity, then $\mathcal{O}^M = I$. In this case, Condition H4 is satisfied, and

$$\bar{g}(u) = \frac{1}{M} \sum_{k=0}^{M-1} \mathcal{O}^{-k-1} g(\mathcal{O}^k u, 0, 0).$$

Of course, if $\mathcal{O} = I$ (i.e., $M = 1$), then $\bar{g}(u) = g(u, 0, 0)$.

With Condition H4, equation (7.1.9) has a solution $U^0(s)$, and it follows from this formal calculation that

$$x_n = P \begin{pmatrix} \mathcal{O}^n U_0(\varepsilon n) \\ \mathcal{S}^n v_0, \end{pmatrix} + O(\varepsilon), \qquad P \begin{pmatrix} U_0(0) \\ v_0 \end{pmatrix} = x_0.$$

The remainder term in this expansion is valid uniformly up to $n = O(1/\varepsilon)$. The proof of this result which we omit, is similar to an analogous result obtained for differential equations (see Hoppensteadt and Miranker, 1976). For a proof in the case where $A = \mathcal{O}(\mathcal{S} = 0)$, see Hoppensteadt and Miranker, 1977.

7.1.4. *The Linear Case*

The linear problem

$$x_{n+1} = (A + \varepsilon B)x_n, \quad x_0 \quad \text{given,}$$

permits us to derive an alternative and more detailed description of the solution of the recurrences which we are studying. In fact it enables us to avoid decomposition of the solution into oscillatory and stable parts as in Section 7.1.3. To proceed we write $x_n = A^n U_n$ and apply the analysis of Section 7.1.3. We find that

$$\bar{g}(u) = A^{-1}\left(\lim_{n \to \infty} \frac{1}{n} \sum_{k=0}^{n-1} A^{-k} B A^k \right) u.$$

If the matrix A is diagonalizable, i.e., there exists an invertible matrix $\mathscr{P}$ such that

$$D \equiv \mathscr{P}^{-1} A \mathscr{P} = (\lambda_k \delta_{kl}),$$

then we may write

$$\bar{g}(u) = A^{-1} \mathscr{P} \left(\lim_{n \to \infty} \frac{1}{n} \sum_{k=0}^{n-1} D^{-k} \mathscr{B} D^k \right) \mathscr{P}^{-1} u, \qquad (7.1.10)$$

where

$$\mathscr{B} \equiv \mathscr{P}^{-1} B \mathscr{P} \equiv (\mathscr{E}_{\mu\nu}).$$

It follows that $D^{-k} \mathscr{B} D^k = (\mathscr{E}_{\mu\nu} \lambda_\mu^{-k} \lambda_\nu^k)$. Thus computing the limit in (7.1.10) leads to consideration of the following expression.

$$\frac{1}{n} \sum_{k=0}^{n-1} (\lambda_\nu/\lambda_\mu)^k = \frac{1}{n} \frac{(\lambda_\nu/\lambda_\mu)^n - 1}{(\lambda_\nu/\lambda_\mu) - 1}.$$

This expression is unity if $\lambda_\mu = \lambda_\nu$, approaches zero if $|\lambda_\nu/\lambda_\mu| \leq 1$ and approaches infinity if $|\lambda_\nu/\lambda_\mu| > 1$. Thus the average exists if and only if the following condition is satisfied.

H5: $\mathscr{E}_{\mu\nu} = 0$ whenever $|\lambda_\nu/\lambda_\mu| > 1$. In this case, we have

$$\lim_{n \to \infty} \frac{1}{n} \sum_{k=0}^{n-1} D^{-k} \mathscr{B} D^k = (\mathscr{E}_{\mu\nu} \delta_{\lambda_\mu, \lambda_\nu}) \equiv \bar{\mathscr{B}}.$$

With Condition H5, we have that

$$\bar{g}(u) = A^{-1} \bar{B} u \equiv A^{-1} \mathscr{P} \bar{\mathscr{B}} \mathscr{P}^{-1} u.$$

7.1.5. *Additional Applications*

We now apply the method of Section 7.1.3 to an elementary genetics model and then to the Munro–Robbins algorithm.

(i) *A population genetics model*

In a large population of diploid organisms having discrete generations, the genotypes determined by one locus having two alleles A and a, divide the population into three groups of type AA, Aa, and aa, respectively. Suppose that the gene pool carried by this population is in proportion p_n of type A in the nth generation. It follows (see Crow and Kimura, '70) that

$$p_{n+1} = p_n + \frac{p_n(1 - p_n)[(w_{11} - w_{12})p_n + (w_{21} - w_{22})(1 - p_n)]}{w_{11}p_n^2 + 2w_{12}p_n(1 - p_n) + w_{22}(1 - p_n)^2},$$

where w_{11}, w_{12}, and w_{22} are relative fitnesses of the genotypes AA, Aa, and aa, respectively.

If the selective pressures act slowly, i.e., if $w_{11} = 1 + \varepsilon\alpha$, $w_{12} = 1$, $w_{22} = 1 + \varepsilon\beta$, where ε is near zero, then

$$p_{n+1} = p_n + \frac{\varepsilon p_n(1 - p_n)[(\alpha - \beta)p_n + \beta]}{1 + O(\varepsilon)}.$$

It follows from the calculation in Section 7.1.3 that

$$p_n = P(\varepsilon n) + O(\varepsilon),$$

where

$$dP/ds = P(1 - P)[(\alpha - \beta)P + \beta].$$

This equation describes the evolution of the A-gene frequency on the slow time scale.

(ii) *Regression analysis*

The Robbins–Munro algorithm (see Robbins and Munro, 1951) is a recurrence relation for approximating the root of a function when only noisy observations of the function can be made.

Let $g(w)$ be a function with a unique root $\hat{w}$ and such $g(w)(w - \hat{w}) > 0$, $w \neq \hat{w}$. Let z_k, $k = 0, 1, \ldots$, be a sequence of identically distributed random variables with mean zero and unit variance. Consider the following recurrence, the Robbins–Munro algorithm, for approximating $\hat{w}$.

$$w(k + 1) = w(k) - \varepsilon\alpha_k[g(w(k)) + \sigma z_k].$$

Here σ and the $\alpha_k, k = 0, 1, \ldots$ are scalars with properties to be specified. We view $g(w(k)) + \sigma z_k$ as a noisy measurement $g(w(k))$. We may expect a very chaotic behavior for the $w(k)$, but we can use the multi-time technique to describe the slow time behavior $w(k)$ which turns out to be quite orderly.

Let's $= \varepsilon k$ and let

$$w(k) = W(k, s) = W_0(k, s) + \varepsilon W_1(k, s) + \ldots .$$

Then by familiar techniques, we find that

$$W_0(k + 1, s) = W_0(k, s),$$

(i.e., W_0 is independent of k and is the slowly varying term being sought) and

$$W_1(k + 1, s) + \frac{dW_0(s)}{ds} = W_1(k, s) - \alpha_k [g(W_0(s)) + \sigma z_k].$$

This equation is solved to give

$$W_1(k, s) = W_1(s) - \sum_{j=0}^{k} a_j [g(W_0(s)) + \sigma z_j] - k \frac{dW_0}{ds}.$$

From this, the usual multi-time hypothesis gives

$$\frac{dW_0}{ds} = - \lim_{k \to \infty} \left[\frac{1}{k} \sum_{j=0}^{k} \alpha_j \right] g(W_0) - \lim_{k \to \infty} \frac{\sigma}{k} \sum_{j=0}^{k} \alpha_j z_j. \qquad (7.1.11)$$

We suppose that the following two averages exist.

$$\bar{\sigma}^2 = \lim_{k \to \infty} k^{-2} \sum_{j=0}^{k} \alpha_j^2,$$

$$\bar{\alpha} = \lim_{k \to \infty} \frac{1}{k} \sum_{j=0}^{k} \alpha_j.$$

Then using the central limit theorem, we conclude that

$$\lim_{k \to \infty} \frac{\sigma}{k} \sum_{j=0}^{k} \alpha_j z_j = N(0, \bar{\sigma}^2).$$

If we further assume that $\bar{\sigma} = 0, (7.1.11)$ implies that

$$dW_0/ds = - \bar{\alpha} g(W_0).$$

For appropriate g, the solutions of this differential equation converge to the equilibrium value $\hat{w}$, and the theory presented here leads us to conclude that

$$w(k) = W_0(s)(1 + O(\varepsilon)).$$

7.2. SINGULARLY PERTURBED BOUNDARY VALUE PROBLEMS

7.2.1. Introduction

While this monograph deals with the initial value problem, we conclude it in this section with a discussion of the numerical treatment of a *singularly perturbed boundary value problem*. We do this in order to point out the existence of a large and difficult class of stiff problems. Boundary value problems certainly find wide applications and computational treatment, as is well known. Along with initial value problems, the stiff cases seriously defy numerical treatment. We refer to Hemker and Miller, 1979 for a representative collection of problems and methodology and also to Il'in, 1969, Pearson, 1968, Dorr, 1971 and Abrahamson, Keller and Kreiss, 1974, for representative earlier work.

We treat the boundary value problem taken in the following canonical form.

$$My \equiv \varepsilon y'' + f(x)y' + g(x)y = h(x), \quad 0 < x < 1,$$
$$y(0) = \alpha, y(1) = \beta, \tag{7.2.1}$$

where ε is considered to be small. We restrict our attention to this linear case for reasons of convenience. The boundary value problem is further complicated by the presence of points where f vanishes; the so-called *turning points*.

7.2.2. Numerically Exploitable Form of the Connection Theory

The algorithm developed here consists of an elaborate discretization of the *connection theory*, the latter describing a rich and ramified structure for solutions of (7.2.1). This structure transcends the comparatively simple fast and slow mode picture of the initial value problem.

We begin by characterizing the analytical form of the solution of (7.2.1) (i.e., development of the WKB or connection theory, (see O'Malley, 1974)) but in a numerically exploitable form (see Miranker and Morreeuw, 1974). This will then be combined with boundary layer methods. As in the context of initial value problems, this approach furnishes numerical methods which inherit the favorable features of the analytic methods;

improvement rather than degradation with increasing stiffness (decreasing ε).

We require some terminology which we introduce by means of the following definitions and remark. In these definitions, all points and sets lie in the interval $[0, 1]$.

DEFINITION 7.2.1. A point x is said to be an *irregular point* if in every neighborhood of x, the function $f(x)$ is neither larger than a positive number nor smaller than a negative number.

REMARK 7.2.2. Turning points are irregular points.

DEFINITION 7.2.3. An *interval of regularity* is an interval containing no irregular points.

DEFINITION 7.2.4. A *neighborhood of irregularity* is an open interval containing exactly one irregular point.

DEFINITION 7.2.5. A right-(left-) sided neighborhood of irregularity is an open interval containing no point of irregularity and whose greatest lower (least upper) bound is a point of irregularity. When we need not specify the right- or left-sidedness, we will refer to those neighborhoods as *demi-neighborhoods of irregularity*.

Form of the solution in an interval of regularity
In a closed interval of regularity, the solution y of (7.2.1) is written in the form

$$y = u + v, \tag{7.2.2}$$

where

(a) $\quad v = e^{-\phi/\varepsilon} w,$
(b) $\quad \phi' = f,$
(c) $\quad Mu = h,$ (7.2.3)
(d) $\quad Mv = 0.$

From these, in turn, we obtain the following equation for w.

(e) $\quad \varepsilon w'' - d(fw)/dx + gw = 0.$

Introducing the operator L and its adjoint L^*:

$$Lz \equiv f\,dz/dx + gz,$$
$$L^*z \equiv -\,d(fz)/dx + gz,$$

we may write (7.2.3c) and (7.2.3d), respectively, as

(a) $\varepsilon u'' + Lu = h,$
(b) $\varepsilon w'' + L^*w = 0.$

In fact, (7.2.3e) may be written as

(c) $M^*w = 0,$

where M^* is the adjoint of M.

Form of the solution in a neighborhood of irregularity
We restrict our attention to the case in which f has nonvanishing right-
and left-sided derivatives at the irregular point, hereafter denoted by x_*. f
may be written as

$$f(x) = a(x - x_*)\left[1 + \frac{(x - x_*)}{2a}f''(x_* + \theta(x - x_*))\right],$$

where

$$a = f'(x_*).$$

We introduce the new variable η in place of x through

$$\eta = \eta(x) = \left[\frac{2}{a}\int_{x_*}^{x} f(s)\,ds\right]^{1/2},$$

where

$$(x - x_*)\eta(x) \geq 0.$$

Note that $\eta(x_*) = 0$, $\eta'(x_*) = 1$ and that $\eta' > 0$ in the demi-neighborhood.
Thus, the change of variables is a valid one, and from (7.2.1), we obtain

$$\varepsilon y_{\eta\eta} + \left(a\eta - \frac{\varepsilon\eta''}{(\eta')^2}\right)y_\eta + a^2\frac{\eta^2}{x^2}\frac{g(x)}{(f'(x))^2}y = \frac{h(x)}{(\eta')^2}. \qquad (7.2.4)$$

The solutions of (7.2.4) are characterized by the following theorem.

THEOREM 7.2.6. *There exist functions* $M(\eta, \varepsilon)$, $N(\eta, \varepsilon)$, $\tilde{h}(\eta, \varepsilon)$ *and* $\sigma(\varepsilon)$, *analytic in ε and continuously differentiable in η such that*

$$y(\eta, \varepsilon) = M(\eta, \varepsilon)z + \varepsilon N(\eta, \varepsilon)z_\eta \tag{7.2.5}$$

is a solution of (7.2.4), where z is a solution of

$$\varepsilon z_{\eta\eta} + a\eta z_\eta + (b + \varepsilon\sigma(\varepsilon))z = \tilde{h}(\eta, \varepsilon). \tag{7.2.6}$$

Here and hereafter

$$b = g(x_*).$$

Proof. Introduce $\zeta(\eta)$ and $\theta(\eta)$ as follows.

(a) $\quad \zeta(\eta) = -\dfrac{\eta''}{(\eta')^2}$,

(b) $\quad \theta(\eta) = \dfrac{a^2\eta^2 g(x) - x^2(f'(x))^2 b}{\eta x^2 (f'(x))^2}$.

Then (7.2.4) may be written as

$$\varepsilon y_{\eta\eta} + (a\eta + \varepsilon\zeta(\eta))y_\eta + (b + y\theta(\eta))y = h(x)/(\eta')^2. \tag{7.2.7}$$

Inserting (7.2.5) into (7.2.7) and using (7.2.6), we get

$$Az + \varepsilon Bz_\eta + C = 0,$$

where

(a) $\quad A = \eta(aM_\eta + \theta M) + \varepsilon(M_{\eta\eta} + \zeta M_\eta - \sigma M - (b + \varepsilon\sigma)2N_\eta + \zeta N))$,

(b) $\quad B = 2M_\eta - (Na\eta)_\eta + \theta\eta N + \zeta(M - Na\eta) + \varepsilon(N_{\eta\eta} + \zeta N_\eta - \sigma N)$,

(c) $\quad C = M\tilde{h} - h + \varepsilon(N_\eta\tilde{h} + \zeta N\tilde{h} + (N\tilde{h})_\eta)$. $\tag{7.2.8}$

Set

$$M = \sum_{i=0}^{\infty} \varepsilon^i M_i \quad \text{and} \quad N = \sum_{i=0}^{\infty} \varepsilon^i N_i.$$

Then the terms of order zero in ε in (7.2.8a) and (7.2.8b) yield

(a) $\quad aM_{0,\eta} + \theta M_0 = 0$,

(b) $\quad 2M_{0,\eta} - (N_0 a\eta)_\eta + \theta\eta N_0 + \zeta(M_0 - aN_0\eta) = 0$. $\tag{7.2.9}$

Let M_0 be the solution of (7.2.9a) which satisfies the condition $M_0(0) = 1$. Then the solution of (7.2.9b), which is bounded at $\eta = 0$, is given by

$$aN_0\eta = M_0 - \exp\int_0^\eta \left(\frac{\theta}{a} - \zeta\right)d\eta.$$

Similarly, to order i in ε, the equations (7.2.8) give

(a) $\quad \eta(aM_{i,\eta} + \theta M_i) - \sigma_{i-1} M_0 + K_{i-1}(\eta) = 0,$ $\qquad$ (7.2.10)

(b) $\quad -(N_i a\eta)' + (\theta/a)a\eta N_i - \zeta a\eta N_i + 2M_{i,\eta} + \zeta M_i + J_{i-1} = 0.$

Here, K_{i-1} which depends on $M_0, \ldots, M_{i-1}, N_0, \ldots, N_{i-1}, \sigma_0, \ldots, \sigma_{i-2}$ is continuous at $\eta = 0$. J_{i-1} depends on $N_0, \ldots, N_{i-1}, \sigma_0, \ldots, \sigma_{i-2}$.

By choosing $\sigma_{i-1} = K_{i-1}(0)$, (7.2.10a) may be solved for M_i which is continuous at $\eta = 0$. With this M_i, (7.2.10b) may be solved in turn for $N_i a\eta$ with N_i being bounded at $\eta = 0$. In this manner, M and N may be constructed. Similarly, $\tilde{h} = \sum_{i=0}^{\infty} \tilde{h}_i \varepsilon^i$ may also be obtained, completing the proof of the theorem. $\qquad\qquad\qquad\qquad\qquad\qquad\qquad\qquad\qquad\qquad$ $\square$

If z is a solution of (7.2.6) such that εz_η is bounded, then we conclude from (7.2.5) that

$$y = M_0 z + \varepsilon N_0 z_\eta + O(\varepsilon). \qquad (7.2.11a)$$

Differentiating (7.2.5) with respect to η and using (7.2.6) and the boundedness of εz_η gives

$$y_\eta = (M_0 - N_0 a\eta) z_\eta + O(1). \qquad (7.2.11b)$$

From (7.2.11) it is clear that if we restrict attention to quantities determined up to $O(\varepsilon)$, then we may use z, which is obtained from the restriction of (7.2.6) to

$$\varepsilon z_{\eta\eta} + a\eta z_\eta + bz = \tilde{h}_0(\eta). \qquad (7.2.12)$$

This equation has a solution of the form $\tilde{h}_0(0)/b + O(\eta) + O(\varepsilon)$, whose derivative moreover, is bounded with respect to both η and ε. Thus, all bounded solutions y of the original equation (7.2.1) may be written in the following form.

(a) $\quad y = M_0(z + h(x_*)/g(x_*)) + \varepsilon N_0 z_\eta + O(\varepsilon, \eta),$

(b) $\quad y_\eta = (M_0 - N_0 a\eta) z_\eta + O(1)_{(\varepsilon,\eta)}.$ $\qquad$ (7.2.13)

Here z is a bounded solution of

$$\varepsilon z_{\eta\eta} + a\eta z_\eta + bz = 0, \qquad (7.2.14)$$

i.e., the homogeneous equation (7.2.12).

In (7.2.13) and hereafter, we use the following notation.

$$f = O(x, y) \Rightarrow |f|/(|x| + |y|) < \text{const} \quad \text{for} \quad (|x| + |y|) < \text{const}.$$

Likewise

$$f = O(1)_{(x,y)} \Rightarrow |f| < \text{const} \quad \text{for} \quad (|x| + |y|) < \text{const}.$$

For the bounded solution of (7.2.14), it may be verified that ηz_η as well as $\varepsilon z_{\eta\eta}$ are bounded. Thus, since $x - \eta = O(\eta^2)$, we find that $z(x) - z(\eta) = O(\eta)$ and that $\varepsilon z_\eta(x) - \varepsilon z_\eta(\eta) = O(\eta)$. Using these observations and the regularity of M_0 and N_0, we write (7.2.13) as follows.

(a) $y(x) = M_0(z + h(x_*)/g(x_*)) + \varepsilon N_0 z_x + O(\varepsilon, x - x_*),$

(b) $y'(x) = (M_0 - N_0 a(x - x_*))z_x + O(1)_{(\varepsilon,x)}.$ (7.2.15)

We adopt the normalization $M_0(x_*) = 1$. Thus, (7.2.15) is further simplified to

(a) $y = z + \dfrac{h(x_*)}{g(x_*)} + \varepsilon N_0(0)z' + O(\varepsilon, x - x_*),$

(b) $y' = z' + O(1)_{(\varepsilon, x - x_*)},$ (7.2.16)

where z is a solution of

$$\varepsilon z'' + a(x - x_*)z' + bz = 0.$$ (7.2.17)

We are now directed to the solutions of (7.2.17). These are the *parabolic cylinder functions*. We now summarize the properties of these solutions which are required for the numerical method.

In Table 7.2-1, u_1 and u_2 are approximations to independent solutions of (7.2.17) (which are bounded in neighborhoods of x_*). Here and throughout $p = b/a$. From this table, we can deduce properties of $z + \varepsilon N_0(0)z'$ and z' needed for y and y' in (7.2.16) by taking appropriate combinations of u_1 and u_2. We recombine the entries in Table 7.2-1 calling them Y_i, $i = 1, 2$ where

(a) $Y_i = u_i + \varepsilon N_0(0)u_i',$ $i = 1, 2,$

(b) $Y_i' = u_i',$ $i = 1, 2.$

These quantities are needed for determining y (as in (7.2.16)). The relevant data are displayed in Table 7.2-2 in terms of the original notation.

In Table 7.2-2, w_2 and w_1 are normalized solutions of $Lw = 0$ and $L^*w = 0$, respectively. The normalization is such that there exist constants p_1 and p_2 such that

$$\lim_{x \to x_*} \frac{fw_1}{|x - x_*|^{p_1}} = \lim_{x \to x^*} \frac{w_2}{|x - x_*|^{p_2}} = 1.$$ (7.2.18)

TABLE 7.2-1: Properties of Solutions of (7.2.17)

	u_1	u_2	Restrictions
$u(x)$	$\exp\left[-\dfrac{a}{\varepsilon}\displaystyle\int_{x_*}^{x}\sigma d\sigma\right]\|x-x_*\|^{p-1}$	$\|x-x_*\|^{-p}$	$\dfrac{\|x-x_*\|}{\varepsilon}\gg 1$
$u'(x)$	$-\dfrac{a}{\varepsilon}\exp\left[-\dfrac{a}{\varepsilon}\displaystyle\int_{x_*}^{x}\sigma d\sigma\right]\mathrm{sig}(x-x_*)\|x-x_*\|^{p}$	$-\mathrm{sig}(x-x_*)p\|x-x_*\|^{-p-1}$	$\dfrac{\|x-x_*\|}{\varepsilon}\gg 1$
$u(x_*)$	$2^{(p-1)/2}\dfrac{\Gamma(1/2)}{\Gamma((1-p)/2)}\mathrm{Re}\left(\left(\dfrac{a}{\varepsilon}\right)^{(1-p)/2}\right)$	$2^{-p/2}\,\mathrm{Re}\left(\left(-\dfrac{a}{\varepsilon}\right)^{p/2}\right)\dfrac{\Gamma(1/2)}{\Gamma((1+p)/2)}$	
$u'(x_*)$	$\mathrm{sig}(x-x_*)2^{(p-1)/2}\dfrac{\Gamma(1/2)}{\Gamma((1-p)/2)}\mathrm{Re}\left(\left(\dfrac{a}{\varepsilon}\right)^{(1-p)/2}\right)$	$\mathrm{sig}(x-x_*)2^{-(p+1)/2}\dfrac{\Gamma(-1/2)}{\Gamma(p/2)}\mathrm{Re}\left(\left(\dfrac{-a}{\varepsilon}\right)^{(p+1)/2}\right)$	

TABLE 7.2-2: Properties of Solutions of (7.2.17)

	$Y_1(x)$	$Y_2(x)$	Restrictions
$Y(x)$	$\exp\left[-\dfrac{1}{\varepsilon}\int_{x_*}^{x} f(s)ds\right]w_1(x)$	$w_2(x)$	$+O(\varepsilon/(x-x_*), x-x_*)$
$Y(0)$	$\dfrac{\text{sig}(x-x_*)}{2\varepsilon}\dfrac{\Gamma(1/2)}{\Gamma(1-p/2)}\,\text{Re}\left(\left(\dfrac{a}{2\varepsilon}\right)^{-(p+1)/2}\right)$	$\dfrac{\Gamma(1/2)}{\Gamma((1+p)/2)}\,\text{Re}\left(\left(\dfrac{-a}{2\varepsilon}\right)^{p/2}\right)$	
$Y_x(x)$	$-\dfrac{f(x)}{\varepsilon}\exp\left[-\dfrac{1}{\varepsilon}\int_{x_*}^{x} f(s)ds\right]w_1(x)$	$-\dfrac{g(x)}{f(x)}w_2(x)$	$+O(1)$
$Y_x(0)$	$\dfrac{1}{2\varepsilon}\dfrac{\Gamma(-1/2)}{\Gamma((1-p)/2)}\,\text{Re}\left(\left(\dfrac{a}{2\varepsilon}\right)^{-p/2}\right)$	$\dfrac{\Gamma(-1/2)}{\Gamma(p/2)}\,\text{sig}(x-x_*)\,\text{Re}\left(\left(\dfrac{a}{2\varepsilon}\right)^{(p+1)/2}\right)$	

Thus, from (7.2.16), we see that in each demi-neighborhood of irregularity the solution of (7.2.1) is a combination of Y_1 and Y_2, viz.

(a) $y = \dfrac{h(x_*)}{g(x_*)} + \lambda Y_2 + \mu Y_1 + O(\varepsilon/(x - x_*), x - x_*),$

(b) $y' = \mu Y_1 + O(1)_{(\varepsilon/(x - x_*), x - x_*)}.$

$\hspace{8cm}$ (7.2.19)

Thus, we see that up to terms which are $O(1)_{(\varepsilon, x - x_*)}$, Y_1 may be identified with the function v and Y_2 with the function u introduced in (7.2.2). Thus, in an appropriate sense, the values of u and v and their derivatives at an irregular point may be read off from Table 7.2-2.

This concludes the description of the solution of (7.1.1) in a neighborhood of irregularity.

7.2.3. Description of the Algorithm

We now derive the numerical method which is carried out on the mesh points, $x_i, i = 0, \ldots, N$, where

$$0 = x_0 < x_1 < \ldots < x_N = 1.$$

Notice that the mesh points are not necessarily equally spaced.

The mesh points consist of three types: irregular points, neighboring points and regular points. These are specified in the following definitions.

DEFINITION 7.2.7. An *irregular mesh point* is an irregular point in the sense of Definition 7.2.1. We assume that all irregular points are to be found among the mesh points. We also assume that each pair of irregular mesh points are separated by at least two points of the mesh.

DEFINITION 7.2.8. x_i is a *neighboring mesh point* if either x_{i+1} or x_{i-1} are irregular points.

DEFINITION 7.2.9. x_i is a *regular mesh point* if it is neither an irregular mesh point nor a neighboring mesh point.

Hereafter, we drop the qualifying word, mesh, associated with these points, since no confusion will result.

Let $\hat{f}$ denote a *discretization of f*. That is a function which interpolates f on the mesh. Similarly, $\hat{u}$ and $\hat{v}$ will denote discretizations of u and v, respectively. $\hat{\phi}$ will denote a primitive of $\hat{f}$.

In addition to the usual *forward* and *backward divided difference operators*, which will be denoted by a subscript x and $\bar{x}$, respectively, we make use of a *directional divided difference operator*. This is given by

$$a\widehat{\frac{\partial}{\partial x}}(\cdot) = \begin{cases} a(\cdot)_x, & \text{if } a > 0, \\ a(\cdot)_{\bar{x}}, & \text{if } a \le 0. \end{cases}$$

The dual of this operator is given by

$$-\widehat{\frac{\partial}{\partial x}}(a\cdot) = \text{signum}\,(-1)\overline{\frac{\partial}{\partial x}(|a|\cdot)}.$$

(See Dorr, 1971.)

The boundary value problem and the connection theory is now discretized. The resulting discrete problem is solved by an iteration process. We now describe this discretization process for each of the different types of mesh points in turn.

Discretization at a regular point
At a regular point x_i, $Mu = h$ and $Mv = 0$ are discretized respectively, as follows.

$$\varepsilon(\hat{u})_{x\bar{x},i} + f_i\widehat{\frac{\partial}{\partial x}}(\hat{u})_i + g_i\hat{u}_i = h_i \tag{7.2.20}$$

and

$$e^{-\phi_i/\varepsilon}\left[\varepsilon(\hat{v}e^{\phi/\varepsilon})_{x\bar{x},i} - \widehat{\frac{\partial}{\partial x}}(f\hat{v}e^{\phi/\varepsilon})_i + g_i(\hat{v}e^{\phi/\varepsilon})_i\right] = 0. \tag{7.2.21}$$

Since (7.2.21) may be multiplied by a constant, the choice of the primitive $\hat{\phi}$ occurring there is arbitrary. The sign of f determines the utilization of the data as is specified in the following remark.

REMARK 7.2.10. The directionally discretized terms in (7.2.20) and (7.2.21), respectively involve u_i and u_{i+1} and v_i and v_{i-1} or they involve u_i and u_{i-1} and v_i and v_{i+1}, depending on the sign of f.

Treatment of a neighboring point
The principal difficulty at a neighboring point x_i involves the evaluation of the second divided difference. We are ill-advised to use (7.2.20) and (7.2.21) at a neighboring point, since the terms of the form $(\cdot)_{x\bar{x},i}$ occurring there will involve values at both an irregular point and at a neighboring

point; these points delimiting a region of rapid change of u and/or v, when ε is small.

We proceed to obtain an alternative approximation to the second derivatives. Let r denote the second derivative of u. Then (7.2.3c) becomes

$$\varepsilon r + fu' + gu = h. \tag{7.2.22}$$

Differentiating (7.2.22) gives

$$fr + (f' + g)u' + g'u = h' - \varepsilon r'. \tag{7.2.23}$$

Combining (7.2.22) and (7.2.23) gives

$$r[f^2 - \varepsilon(f' + g)] = (gu - h)(f' + g) + f[(h' - \varepsilon r') + g'u]. \tag{7.2.24}$$

If the last bracket here is bounded at x_i, we may neglect it, since its coefficient f is small at x_i (since we assume that $|x_i - x_{i*}|$ is small), where x_{i*} denotes the irregular point for which x_i is a neighbor. Then we have the following approximation for r.

$$\varepsilon r = \chi_1(gu - h), \tag{7.2.25}$$

where

$$\chi_1 = \varepsilon \frac{g + f'}{f^2 - \varepsilon(f' + g)}. \tag{7.2.26}$$

Similarly, denoting the second derivative of $w = \varepsilon^{\phi/\varepsilon}v$ by s, we obtain from (7.2.3d) the following equations in place of (7.2.22), (7.2.23), (7.2.24) and (7.2.26), respectively.

$$\varepsilon s - fw' + (g - f')w = 0,$$
$$-fs + (g - 2f')w' + (g - f')'w = -\varepsilon s'$$

and

$$\varepsilon s = \chi_2(g - f')w,$$

where

$$\chi_2 = \varepsilon \frac{g - 2f'}{f^2 - \varepsilon(g - 2f')}.$$

Using r and s in (7.2.20) and (7.2.21), respectively, in place of the second divided differences, Remark 7.2.10 shows that one of the resulting equations does not make use of data at the irregular point. We use this equation to calculate the associated function (u or w as the case may be) at the neighboring point. We call this function the *principal function* (relative to this neighboring point). The remaining function is called the *minor function*.

Let y denote the principal function (u or $e^{(\phi - \phi_{i\bullet})/\varepsilon}v$, as the case may be), and let $\bar{y}$ denote the minor function.

First consider the case where y is u. Then write (7.7.22) as

$$Ly = h - \varepsilon r. \tag{7.2.27}$$

We solve this equation to obtain

$$y = \lambda + \mu w,$$

where λ is a particular solution of (7.2.27), μ is a constant and w is a normalized solution (in the sense of (7.2.18)) of $Ly = 0$. Writing

$$\hat{y}_i = \lambda + \mu \hat{w}_i, \tag{7.2.28}$$

we approximate λ as follows.

$$\lambda = (h_i - \varepsilon r_i)/g_i. \tag{7.2.29}$$

For $\hat{w}_i$, we have

$$\hat{w}_i = |x_i - x_{i\bullet}|^{-g_i/f_i}. \tag{7.2.30}$$

In (7.2.29), r is computed by means of (7.2.25) and (7.2.26), and in (7.2.25), we set u to equal to $\hat{y}_i$ which is known to us. Since y is the principal function $\hat{y}_i$ is known, and so then from (7.2.28)–(7.2.30), μ is known.

In the case that the principal function y is identified with $e^{(\phi - \phi_{i\bullet})/\varepsilon}v$, we similarly derive (7.2.28). However in this case,

$$\lambda = -\varepsilon s_i/g_i \tag{7.2.31}$$

approximately, and $\hat{w}_i$ is obtained from the normalized solution of $L^*y = 0$:

$$\hat{w}_i = (1/f_i)|x_i - x_{i\bullet}|^{g_i/f_i}. \tag{7.2.32}$$

For the minor function, the same development may be made, representing it as

$$\bar{y} = \bar{\lambda} + \bar{\mu}\bar{w}.$$

$\bar{\lambda}$ is computed as in (7.2.30), however, since $\bar{y}$ is not known, we use a prior value of $\bar{y}$ (e.g., $\bar{y}$ from the previous iteration of the solution process referred to for the discretized problem being developed). The equations

(7.2.28)–(7.2.32) are valid for the minor function with bars inserted as necessary. However, because $\bar{y}$ is the minor function, $\bar{y}_i$ is not known, and so $\bar{\mu}$ is unknown as well. To determine $\bar{y}_i$, we must determine $\bar{\mu}$, which we now proceed to do.

Treatment at an irregular point

The study of the principal function at neighboring points permits us to express this function as

$$\lambda_- + \mu_- \, w_-,$$

to the left of an irregular point and as

$$\lambda_+ + \mu_+ \, w_+,$$

to the right. Similarly, the minor function, to the left of and to the right of an irregular point, may be written as

$$\bar{\lambda}_- + \bar{\mu}_- \bar{w}_- \quad and \quad \bar{\lambda}_+ + \bar{\mu}_+ \bar{w}_+,$$

respectively.

We may identify $\lambda_+ + \bar{\lambda}_+ + \mu_+ w_+ + \bar{\mu}_+ \bar{w}_+$ with (7.2.19a) as an approximation to the solution of y, to one side of an irregular point. From Table 7.2-2, we see that the limiting values as $x_i \to x_{i*}$, of $Y_1(x)$ and $Y_2(x)$ are known and approximate the normalized solutions (at x_{i_*}) which are w_+ and $\bar{w}_+$ in some order. Call Y_1 and Y_2 the functions Y and $\bar{Y}$, as the case may be. Thus, by the continuity of y at an irregular point, we have

$$\lambda_- + \bar{\lambda}_- + \mu_- Y_-(0) + \bar{\mu}_- \bar{Y}_-(0) = \lambda_+ + \bar{\lambda}_+ + \mu_+ Y_+(0) \\ + \bar{\mu}_+ \bar{Y}_+(0). \qquad (7.2.33)$$

Similarly, from (7.2.19b) and the continuity of y' at an irregular point, we have

$$\mu_- Y'_-(0) + \bar{\mu}_- \bar{Y}'_-(0) = \mu_+ Y'_+(0) + \bar{\mu}_+ \bar{Y}'_+(0). \qquad (7.2.34)$$

(7.2.33) and (7.2.34) form a system for the determination of $\bar{\mu}_+$ and $\bar{\mu}_-$. Except in the case where the determinant of this system vanishes (an analogue of a resonance phenomena in the solution), we may solve this system for $\bar{\mu}_+$ and $\bar{\mu}_-$.

7.4.4. Computational Experiments

Computational experimentation with the semianalytic methods developed here require judicious computational technique, and we refer to Section 4

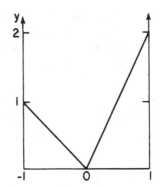

Fig. 7.2-1. Approximate solution in case (ii)

of Miranker and Morreeuw, 1974 for a description of the latter. Here we describe results of such experiments. First we characterize the errors arising through use of the algorithm for the model (test) equation $\varepsilon y'' + axy' + by = 0$ (compare (2.1.2)). Following this, we give several examples of solutions for linear equations with variable coefficients which illustrate types of solution behavior at turning point problems.

Solution of the boundary value problem $\varepsilon y'' + axy' + by = 0$ *with* $y(-1) = 1$,
    $y(1) = 2$.

We take $\varepsilon = 10^{-7}$. Two typical cases are

(i) $a = -1, b < 0$,

(ii) $a = 1, b = -1.0001(\sim -1)$.

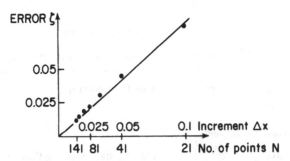

Fig. 7.2-2. Maximum error versus Δx in case (ii)

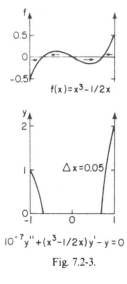

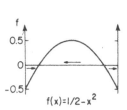

$$10^{-7}y'' + (x^3 - 1/2x)y' - y = 0$$

Fig. 7.2-3.

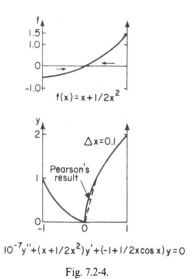

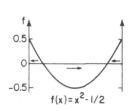

$$10^{-7}y'' + (x + 1/2x^2)y' + (-1 + 1/2x\cos x)y = 0$$

Fig. 7.2-4.

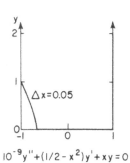

$$10^{-9}y'' + (1/2 - x^2)y' + xy = 0$$

Fig. 7.2-5.

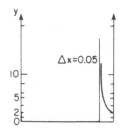

$$10^{-9}y'' + (x^2 - 1/2)y' + xy = 0$$

Fig. 7.2-6.

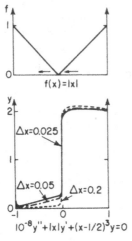

$$10^{-8}y'' + |x|y' + (x-1/2)^3 y = 0$$

Fig. 7.2-7.

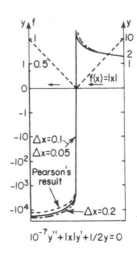

$$10^{-7}y'' + |x|y' + 1/2\,y = 0$$

Fig. 7.2-8.

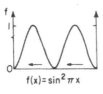

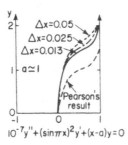

$$10^{-7}y'' + (\sin\pi x)^2 y' + (x-a)y = 0$$

Fig. 7.2-9.

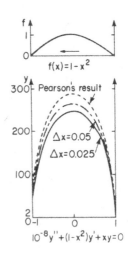

$$10^{-8}y'' + (1-x^2)y' + xy = 0$$

Fig. 7.2-10.

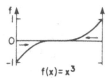

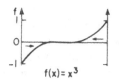

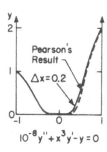

$$10^{-8}y'' + x^3 y' - y = 0$$

Fig. 7.2-11.

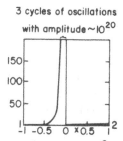

3 cycles of oscillations
with amplitude ~10^{20}

Pearson's Result for $10^{-8}y'' + x^3 y' - xy = 0$
In this case, our method is unstable.

Fig. 7.2-12.

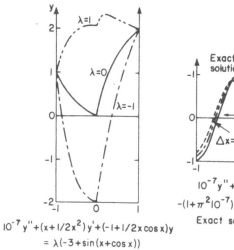

$$10^{-7}y'' + (x + 1/2 x^2)y' + (-1 + 1/2 x \cos x)y$$
$$= \lambda(-3 + \sin(x + \cos x))$$

Fig. 7.2-13.

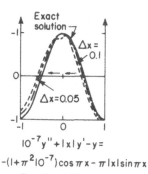

$$10^{-7}y'' + |x| y' - y =$$
$$-(1 + \pi^2 10^{-7}) \cos \pi x - \pi |x| \sin \pi x$$

Exact solution: $y = \cos \pi x$

Fig. 7.2-14.

In Case (i), the algorithm gives zero at all interior mesh points if b is not a (machine) integer. This corresponds to an error of less than 10^{-8}.

In Case (ii), the exact solution is linear to within 10^{-4} on each side of the turning point, where it vanishes. The approximate solution is displayed in Figure 7.2-1. The maximum error ζ versus the mesh increment Δx is plotted in Figure 7.2-2. The linearity of the plot in Figure 7.2-2 is expected since all of the discretization procedures in the various aspects of the algorithm are of the first order in Δx.

Variable coefficients cases

Now we present a sequence of Figures 7.2-3 – 7.2-16 each of which exhibits a particular coefficient $f(x)$ of the term y', the specific choice of the boundary value problem and the corresponding numerical solution. In the plots of $f(x)$, the arrows drawn along the x-axis represent the direction of the discretization of the derivative (see (7.2.20) and (7.2.21)). The various values of the mesh increment Δx employed are also displayed in the figures. The problems treated in the figures have been chosen to illustrate the large variety of possible solution types. Pearson, 1968 has studied some of these cases by a *mesh refinement method*, and we plot his results when there is disagreement with those of the algorithm discussed here. We also compare the numerical results to the exact solution in

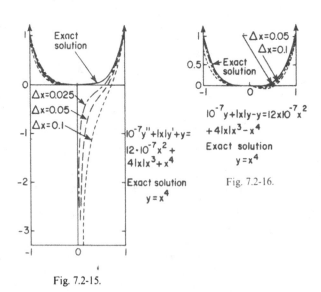

Fig. 7.2-15.

Fig. 7.2-16.

Figures 7.2-14–7.2-16. In these cases, the numerical error vanishes with the mesh increment, and when the *maximum principle* is satisfied by the solution of the boundary value problem, the numerical error is less than the mesh increment.

The sign ($\mp$) of the linear term y in the differential equation determines whether or not the maximum principle holds for the corresponding boundary value problem. The corresponding cases are displayed by Figures 7.2-15 and 7.2-16. Notice the extreme dependence of the computational methods discussed here to the presence or absence of the maximum principle.

References

Abrahamson, L. R., Keller, H. B. and Kreiss, H. O.: (1974), 'Difference Approximations for Singular Perturbations of Systems of Ordinary Differential Equations,' *Num. Math.* **22**, 367–391.

Aiken, R. C. and Lapidus, L.: (1974), 'An Effective Numerical Integration Method for Typical Stiff Systems, *AICHE J*, **20**, 368–374.

Achieser, N. I. and Glassman, I. M.: (1954), *Theorie der Linearen Operatoren in Hilbert Raum*, Akademie-Verlag, Berlin.

Axelsson, O., Frenk, L. and vander Sluis, A., (Eds): (1980), Proceedings of the Conference on Analytical and Numerical Approaches to Asymptotic Problems in Analysis, North-Holland, Amsterdam.

Auslander, L. and Miranker, W. L.: (1979), 'Algebraic Methods in the Study of Stiff Differential Equations, in P. W. Hemker and J. J. H. Miller (Eds.), *Numerical Analysis for Singular Perturbation Problems*, Academic Press, London, pp. 3–23.

Bellman, B.: (1960), *Introduction to Matrix Analysis*, McGraw Hill, New York.

Bjurel, G., Dahlquist, G., Lindberg, B., Linde, S. and Oden, L.: (1972), 'Survey of Stiff Ordinary Differential Equations', Report NA 70.11, Department of Information Processing and Computer Science, The Royal Institute of Technology, Stockholm, Sweden.

Butcher, J. C.: (1964), 'Implicit Runge–Kutta Processes,' *Math. Comp.* **18**, 50–64.

Calahan, D. A.: (1967), 'Numerical Solution of Linear Systems with Widely Separated Time Constants', *Proc. IEEE* **55**, 2016–2017.

Certaine, J.: (1960), 'The Solution of Ordinary Differential Equations with Large Time Constants', in A. Ralston and H. S. Wilf (Eds.), *Mathematical Methods for Digital Computers*, Wiley, New York, pp. 128–132.

Chevelley, C.: (1964), *Theory of Lie Groups*, Princeton Press.

Crow, J. F. and Kimura, M.: (1970), *An Introduction to Population Genetics Theory*, Harper Row, New York.

Curtiss, C. F. and Hirschfelder, J. O.: (1952), 'Integration of Stiff Equations', *Proc. Nat. Acad. Sci. U.S.* **38**, 235–243.

Dahlquist, G.: (1963), 'A Special Stability Problem for Linear Multistep Methods', *BIT* **3**, 27–43.

Dahlquist, G., Bjork, A. and Anderson, N.: (1974), *Numerical Methods*, Prentice-Hall, Englewood Cliffs, New Jersey.

Dorr, F. W.: (1971), 'An Example of Ill-Conditioning in the Numerical Solution of Singular Perturbation Problems,' *Math. Comp.* **25**, 271–283.

Gear, C. W.: (1969), 'The Automatic Integration of Stiff Ordinary Differential Equations', in A. J. H. Morrel (Ed.), *Information Processing*, **68** North-Holland Publ. Co., Amsterdam, pp. 187–193.

Gear, C. W.: (1971), *Numerical Initial Value Problems in Ordinary Differential Equations*, Prentice-Hall, Englewood Cliffs, New Jersey.

Greenberg, H. J. and Konheim, A. G.: (1964), 'Linear and Nonlinear Methods in Pattern Classification', *IBM J. Res. Dev.* **8**, 299–307.

Hemker, D. W. and Miller, J.J.: (Eds.) (1979), *Numerical Analysis of Singular Perturbation Problems*, Academic Press, London.

Henrici, P.: (1962), *Discrete Variable Methods in Ordinary Differential Equations*, Wiley, New York.

Hindmarsh, A. C.: (1974), 'Gear's Ordinary Differential Equation Solver', *UCID-30001 (Rev. 3)*, Lawrence Livermore Lab., Livermore, CA 94550.

Hochshild, G.: (1965), *The Structure of Lie Groups*, Holden-Day, San Francisco.

Hoppensteadt, F. C.: (1971), 'Properties of Solutions of Ordinary Differential Equations with Small Parameters', *Comm. Pure and Appl. Math.* **XXIV**, 807–840.

Hoppensteadt, F. C. and Miranker, W. L.: (1976), 'Differential Equations having Rapidly Changing Solutions: Analytic Methods for Weakly Nonlinear Systems', *J. Differential Equations* **22**, 237–249.

Hoppensteadt, F. C. and Miranker, W. L.: (1977), 'Multi-time Methods for Systems of Differences Equations', *Studies in Appl. Math.* **56**, 273–289.

Il'in, A. M.: (1969), 'A Difference Scheme for a Differential Equation with a Small Parameter Multiplying the Highest Derivative', *Mat. Zametki* **6**, 237–248. *Math. Notes* **6**, 596–602.

Isaacson, E. and Keller, H. B.: (1966), *Analysis of Numerical Methods*, Wiley, New York.

Jain, R. K.: (1972), 'Some *A*-stable Methods for Stiff Ordinary Differential Equations', *Math. Comp.* **26**, 71–78.

Keller, H. B.: (1975), 'Numerical Solution of Boundary Value Problems for Ordinary Differential Equations: Survey and Some Recent Results on Difference Methods', in A. K. Aziz (Ed.), *Numerical Solutions of Boundary Value Problems for Ordinary Differential Equations*, Academic Press, New York.

Lambert, J. C.: (1973), *Computational Methods in Ordinary Differential Equations*, Wiley, London.

Leveque, W.: (1956), *Topics in Number Theory*, Addison–Wesley, Reading, MA.

Levin, J. and Levinson, N.: (1954), 'Singular Perturbations of Nonlinear Systems of Differential Equations and Associated Boundary Layer Equation', *J. Rational Mech. Anal.* **3**, 247–270.

Liniger, W.: (1973–74), 'Lecture Notes on Stiff Differential Equations', A course given at the University of Lausanne.

Liniger, W. and Willoughby, R.: (1970), 'Efficient Integration Methods for Stiff Systems of Ordinary Differential Equations', *SIAM J. Numer. Anal.* **7**, 47–66.

Luenberger, D. G.: (1965), *Introduction to Linear and Nonlinear Programming*, Addison–Wesley, Reading, MA.

Micchelli, C. A. and Miranker, W. L.: (1973), 'Optimal Difference Schemes for Linear Initial Value Problems', *SIAM J. Numerical Analysis* **10**, 983–1009.

Micchelli, C. A. and W. L. Miranker, W. L.: (1974), 'Asymptotically Optimal Approximation in Fractional Sobolev Spaces and the Numerical Solution of Differential Equations', *Numer. Math.* **22**, 75–87.

Miranker, W. L.: (1962), 'Singular Perturbation Analysis of the Differential Equations of a Tunnel Diode Circuit', *Q. Appl. Math.* **XX**, 279–299.

Miranker, W. L.: (1962b), 'The Occurrence of Limit Cycles in the Equations of a Tunnel Diode Circuit', *IRE Trans. Circuit Theory* **9**, 316–320.

Miranker, W. L.: (1971), 'Difference Schemes with Best Possible Truncation Error', *Numerische Mathematik* **17**, 124–142.

Miranker, W. L.: (1971b), 'Matricial Difference Schemes for Integrating Stiff Systems of Ordinary Differential Equations', *Math. Comp.* **25**, 717–728.

Miranker, W. L.: (1972), 'Enveloping an Iteration Scheme', *IBM J. Res. Dev.* **16**, 389–392.

Miranker, W. L.: (1973), 'Numerical Methods of Boundary Layer Type for Stiff Systems of Differential Equations', *Computing* **11**, 221–234.

Miranker, W. L.: (1975), 'The Computational Theory of Stiff Differential Equations', *IAC*, Series III-N.102, Rome and No. 219–7667, V. Paris XI, U.E.R. Mathematique, 91405 Orsay, France.

Miranker, W. L. and Hoppensteadt, F.: (1973), 'Numerical Methods for Stiff Systems of Differential Equations Related with Transistors, Tunnel Diodes, etc.', *Lecture Notes in Computer Science* **10**, Springer-Verlag, pp. 416–432.

Miranker, W. L. and Morreeuw, J. P.: (1974), 'Semi-Analytic Numerical Studies of Turning Points Arising in Stiff Boundary Value Problems,' *Math. Comp.* **28**, 1017–1034.

Miranker, W. L. and Wahba, G.: (1976), 'An Averaging Method for the Stiff Highly Oscillatory Problem', *Math. Comp.* **30**, 383–399.

Miranker, W. L., van Veldhuizen, M. and Wahba, G.: (1976), 'Two Methods for the Stiff Highly Oscillatory Problem', in J. Miller (Ed.), *Topics in Numerical Analysis III*, *Proceedings of the Royal Irish Academy Conference on Numer. Anal.*, Academic Press, London.

Miranker, W. L. and van Veldhuizen, M.: (1978), 'The Method of Envelopes', *Math. Comp.* **32**, pp. 453–496.

Miranker, W. L. and Chern, I-L.: (1980), 'Dichotomy and Conjugate Gradients in the Stiff Initial Value Problem', *J. Lin. Alg. Appl.* **30**.

O'Malley, R. E.: (1974), *Introduction to Singular Perturbations*, Academic Press, New York.

Pearson, C. E.: (1968), 'On a Differential Equation of Boundary Layer Type', *J. Math. Phys.* **47**, pp. 134–154.

Persek, S. C. and Hoppensteadt, F. C.: (1978), 'Iterated Averaging Methods for Systems of Ordinary Differential Equations with a Small Parameter', *Comm. Pure Appl. Math.* **XXXI**, 133–156.

Robbins, H. and Munro, S.: (1951), 'A Stochastic Approximation Method', *Ann. Math. Stat.* **22**, 400–407.

Rosenbrock, H. H.: (1962), 'Some General Implicit Processes of the Numerical Solution of Differential Equations', *Comp. J.* **5**, 329–330.

Strang, G.: (1962), 'Trigonometric Polynomials and Difference Methods of Maximum Accuracy', *J. Math. and Phys.* **41**, 147–154.

Snider, A. D.: (1972), 'An Improved Estimate of the Accuracy of Trigonometric Interpolation', *SIAM J. Numer. Anal* **9**, 505–508.

Snider, A. D. and Fleming, G. C.: (1974), 'Approximation by Aliasing with Applications to 'Certaine' Stiff Differential Equations', *Math. Comp.* **28**, 465–473.

Volosov, V. M.: (1962), 'Averaging in Systems of Ordinary Differential Equations', *Russ. Math. Surveys* **17**, 1–126.

Widlund, O.: (1967), 'A Note on Unconditionally Stable Linear Multistep Methods', *BIT* **7**, 65–70.

Willoughby, R. A., (Ed.): (1974), *Stiff Differential Equaqtions*, Proceedings of the IBM Research Symposia Series. Plenum Press, New York.

Index

Mathematics and Its Applications

Managing Editor:

M. HAZEWINKEL
Department of Mathematics, Erasmus University, Rotterdam, The Netherlands

Editorial Board:

R. W. BROCKETT, *Harvard University, Cambridge, Mass., U.S.A.*
J. CORONES, *Iowa State University, Ames, Iowa, U.S.A. and Ames Laboratory, U.S. Department of Energy, Iowa, U.S.A.*
Yu. I. MANIN, *Steklov Institute of Mathematics, Moscow, U.S.S.R.*
G.-C. ROTA, *M.I.T., Cambridge, Mass., U.S.A.*

1. Willem Kuyk, *Complementarity in Mathematics, A First Introduction to the Foundations of Mathematics and Its History.* 1977.
2. Peter H. Sellers, *Combinatorial Complexes, A Mathematical Theory of Algorithms.* 1979.
3. Jacques Chaillou, *Hyperbolic Differential Polynomials and their Singular Perturbations.* 1979.
4. Svtopluk Fučík, *Solvability of Nonlinear Equations and Boundary Value Problems.* Forthcoming.
5. Willard L. Miranker, *Numerical Methods for Stiff Equations and Singular Perturbation Problems.* 1980.
6. P. M. Cohn, *Universal Algebra.* Forthcoming.
7. Vasile I. Istrǎţescu, *Fixed Point Theory, An Introduction.* Forthcoming.